historical geology of north america

historical geology of north america

second edition

Morris S. Petersen
Brigham Young University

J. Keith Rigby
Brigham Young University

Lehi F. Hintze
Brigham Young University

wcb

Wm. C. Brown Company Publishers
Dubuque, Iowa

Consulting Editor

Sherwood D. Tuttle
University of Iowa

Library of Congress Catalog Card Number: 79–52357

ISBN 0–697–05062–9

Printed in the United States of America
10 9 8 7

Contents

Preface

Historical geology, like all areas of geology, has been vastly modified over the past decade as a consequence of the pervasive influence of the concept of plate tectonics. Man's exploration of the moon and nearby planets has also broadened our understanding of earth history, especially the earliest part of that record. The effects of these expanded horizons is incorporated in this second edition of *Historical Geology of North America*.

Historical Geology of North America is intended for use in general education courses, or with supplementary material for standard historical geology classes. Detailed information has been omitted, or only very briefly discussed, in order to present the story of historical geology in a brief, simple fashion. Principles upon which this story is based will require additional explanation by the course instructor. This revised story of the physical and biological evolution of our earth is presented using North America as an example. It is our hope that the account is understandable and enjoyable to the introductory student of historical geology.

The writers gratefully acknowledge the exceptional service of Mr. Art Lee, who served as draftsman for this edition.

MSP
JKR
LFH

1

Present Is the Key to the Past

TOPICS

UNLOCKING THE PAST

Deciphering earth history is a study that demands the efforts of the most talented geologic Sherlock Holmes. The study is at the same time deductive and integrative; success requires the understanding of four concepts: geologic time, paleogeography, and geologic and biologic processes. In these pages we will simplify our task by limiting most of our discussion to North America. We will examine geologic history by proceeding from the most ancient towards the present. The study is exciting because it documents two of geology's greatest contributions to modern thought: the immensity of time and the constancy of change.

UNIFORMITARIANISM

Up until the late 1700s, most students of nature believed that the earth was only a few thousand years old and that its history had been marked by violent worldwide catastrophes, the last of which had been the Biblical flood. In 1785, shortly after the American Revolution, a Scotsman, James Hutton, put forth the concept of uniformitarianism. This idea proposed that the geologic record should be interpreted on the basis that geologic processes operate today as they did in earlier times. Because most geologic processes proceed slowly, reason requires that geologic time be incredibly long to account for the present world. Hutton realized this implication when he wrote that for the earth he saw "no vestige of a beginning, no prospect of an end." Uniformitarianism has been paraphrased, "The present is the key to the past."

Hutton's idea was incorporated into the world's first geological textbook, *Principles of Geology* (1830) by Charles Lyell, and it changed the course of geologic thought. Today we qualify the idea that the past functioned exactly like the present because we have come to realize that in their very early stages, the earth's atmosphere and hydrosphere were different than they are today and therefore would not exactly duplicate the same processes. Also, earth's earliest days were shorter because the earth was rotating more rapidly. Furthermore, the earth's internal source of energy, radioactivity, was more abundant in earlier times and was able to operate the earth's internal processes more vigorously than today. Nevertheless, the same basic laws governing matter and energy have always been in operation. But geologic and biologic processes have changed in rate and relative importance as the earth has evolved to its present condition.

EARTH'S TECTONIC MOTOR

Tectonics is a word that encompasses all deformation of the earth's crust. It is expressed on all scales, ranging from microscopic disruption in rocks, through folds and faults in mountain systems, up to movement of the continents and ocean floors themselves. No part of the earth's surface remains stationary, although the rate of movement is generally so slow that it has taken mankind a long time to recognize just how pervasive tectonic movements really are.

Principally by the study of the way that earthquake shocks are transmitted through and around the earth, we have determined that the earth's interior is layered as shown on figure 1.1. The earth's core, believed to be mostly iron, constitutes about a sixth of the earth's volume and a third of its mass. Fluid movements within the liquid outer core are thought to be responsible for generating the earth's magnetic field. The energy source that drives the outer core's fluid motion is not surely known, but it may be gravitational and thermal energy, or energy derived from the earth's rotation. Throughout geologic time the earth's magnetic field has left its varying imprint on rocks of continents and ocean floors creating a tape recording of tectonic movements of the past.

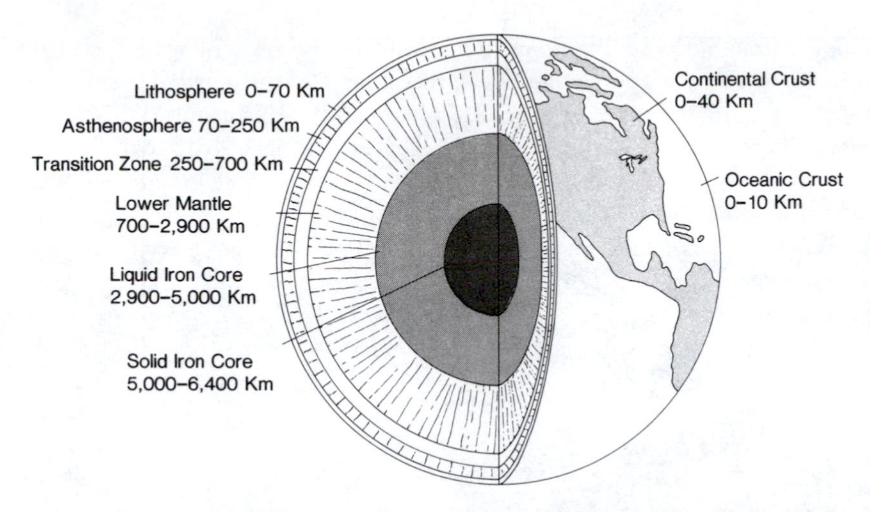

Figure 1.1. Earth's major layers as interpreted chiefly from seismic data.

Above the core the lower mantle forms a solid shell that passes upwards through a transition zone into the asthenosphere, a zone of weakness. The asthenosphere transmits seismic waves at a slower velocity than the adjacent shells, and this is interpreted to mean that it is a zone of partial (perhaps 1 to 10 percent) melting.

The lithosphere is the outer shell of the earth in which continents are embedded. It transmits seismic waves rapidly and consists of strong, solid rock. The lithosphere rests upon a mobile foundation. The asthenosphere, and possibly the transition zone of the outer mantle are in constant slow motion. They appear to behave like hot semisolids that can "flow" at the ponderous rate of about a centimeter a year. The energy sources for this convective movement appear to be heat given off by radioactive decay of potassium, uranium, thorium, and rubidium that were part of the earth's original constituents, and gravity.

The ponderous turbulent movement in the aesthenosphere is transmitted into movement of the outer shell of the earth. This shell breaks apart along midocean ridges called spreading centers, where rising magma from the partially melted zones below freezes onto the oceanic crust. Older oceanic crust is recycled into the earth's interior along oceanic trenches called subduction zones. Thus the earth's lithosphere is broken into plates, large and small, the present pattern of which is shown on figure 1.2.

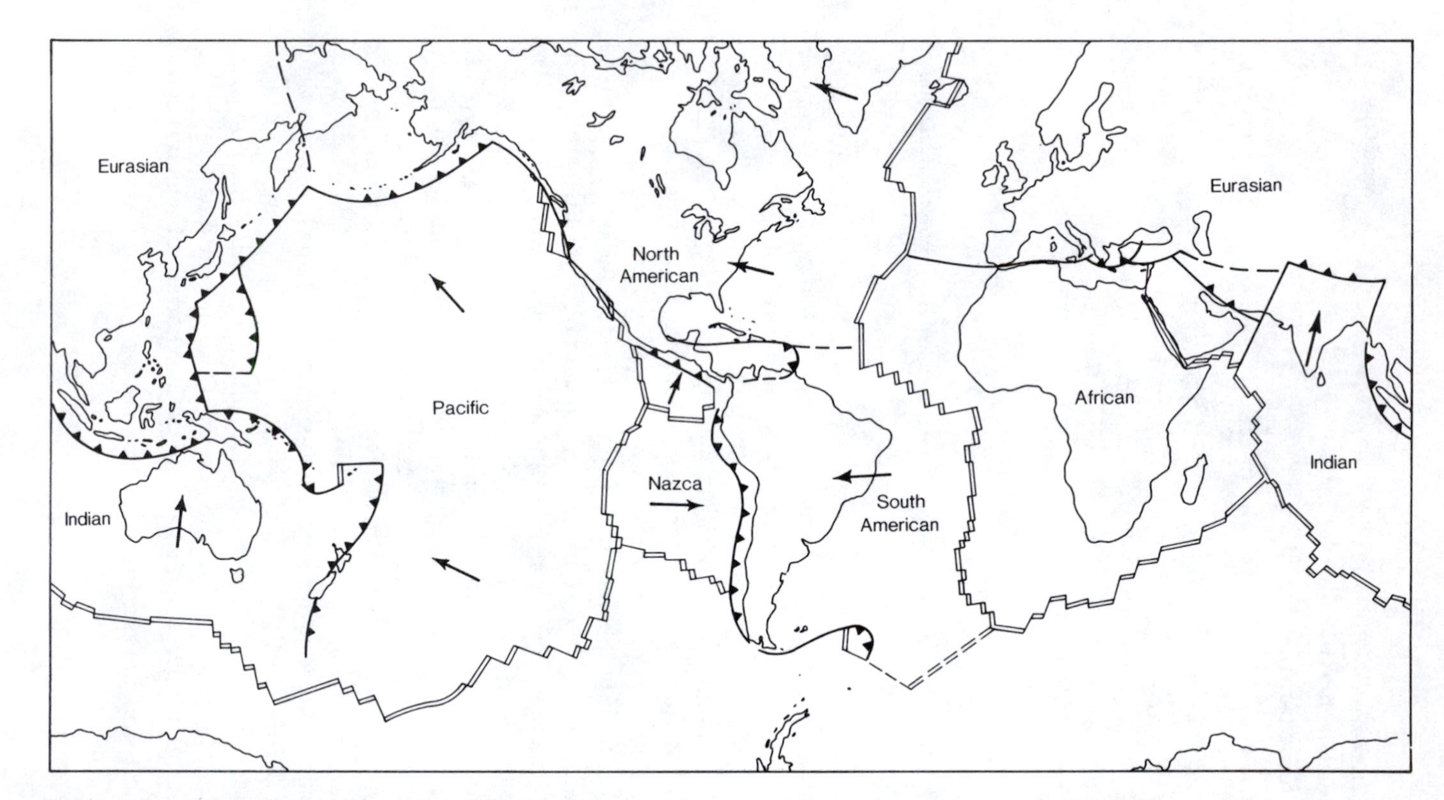

Figure 1.2. Major plates of the present world. Midocean ridge spreading centers shown as double lines. Subduction zones marked by toothed lines, teeth on the upper plate. Transform ("slide-by") boundary faults shown as single lines. Dashed lines indicate weak or uncertain boundaries. Arrows show direction of plate motion.

EARTH'S GEOLOGIC PATTERNS

Earth's present complex geologic patterns are usually represented on geologic maps by a patchwork of jolly colors showing rocks of various ages. To the trained eye this array shows the long-term interplay between the deposition of rocks, mostly as flat layers, and their subsequent deformation by folding and faulting. Continents are more complex than ocean basins because continental materials, being lightweight, have generally not been recycled into the earth's interior, and thus show superimposed imprints of geologic processes over a longer part of earth history. Figure 1.3 shows continental areas divided into four categories. The oldest continental rocks are in the shield areas that formed the nucleus for continental growth. Shields have been around so long (in excess of 600 million years) that they have long since been worn by erosion to lands of low relief, as for example, most of eastern Canada. Old fold belts (200 to 600 million years old) include thick sedimentary and volcanic accumulations that have been deformed into ancient mountain chains, now much worn down. The Appalachian Mountains are a well-known example. Young fold belts include all mountain systems younger than 200 million years, some of them still actively growing as indicated by volcanic and seismic activity. Young fold belts occur along boundaries of major crustal plates, as can be readily seen by comparing figures 1.2 and 1.3. The western cordillera of North and South America are conspicuous examples of young mountain systems. Unshaded areas on figure 1.3 are platforms where nearly horizontal strata cover shield areas or old mountain belts.

Because of the inaccessibility of the deep sea floor, the geologic pattern of ocean basins has only been revealed in recent decades. It is very different than the continental pattern. Ocean basins contain no shield rocks, nor even any as old as the older continental fold belts. Figure 1.4 shows the age of the ocean floors as it has been deduced from deep-sea drilling and studies of the magnetic patterns of sea floor rocks. Compared to the complexities of the continents, the sea floor patterns are very simple. New magmatic rock is generated along midocean spreading centers. As time goes on this rock moves away from the spreading center on symmetrically diverging conveyor belts. Where there is a subduction zone at the oceanic margin, the old ocean floor is consumed by being subducted into the earth's interior. This is what is happening in the western Pacific Ocean, where the oldest remaining ocean floor is that part farthest from the Pacific spreading center. Earthquakes are associated with both crustal generation and crustal subduction, those linked

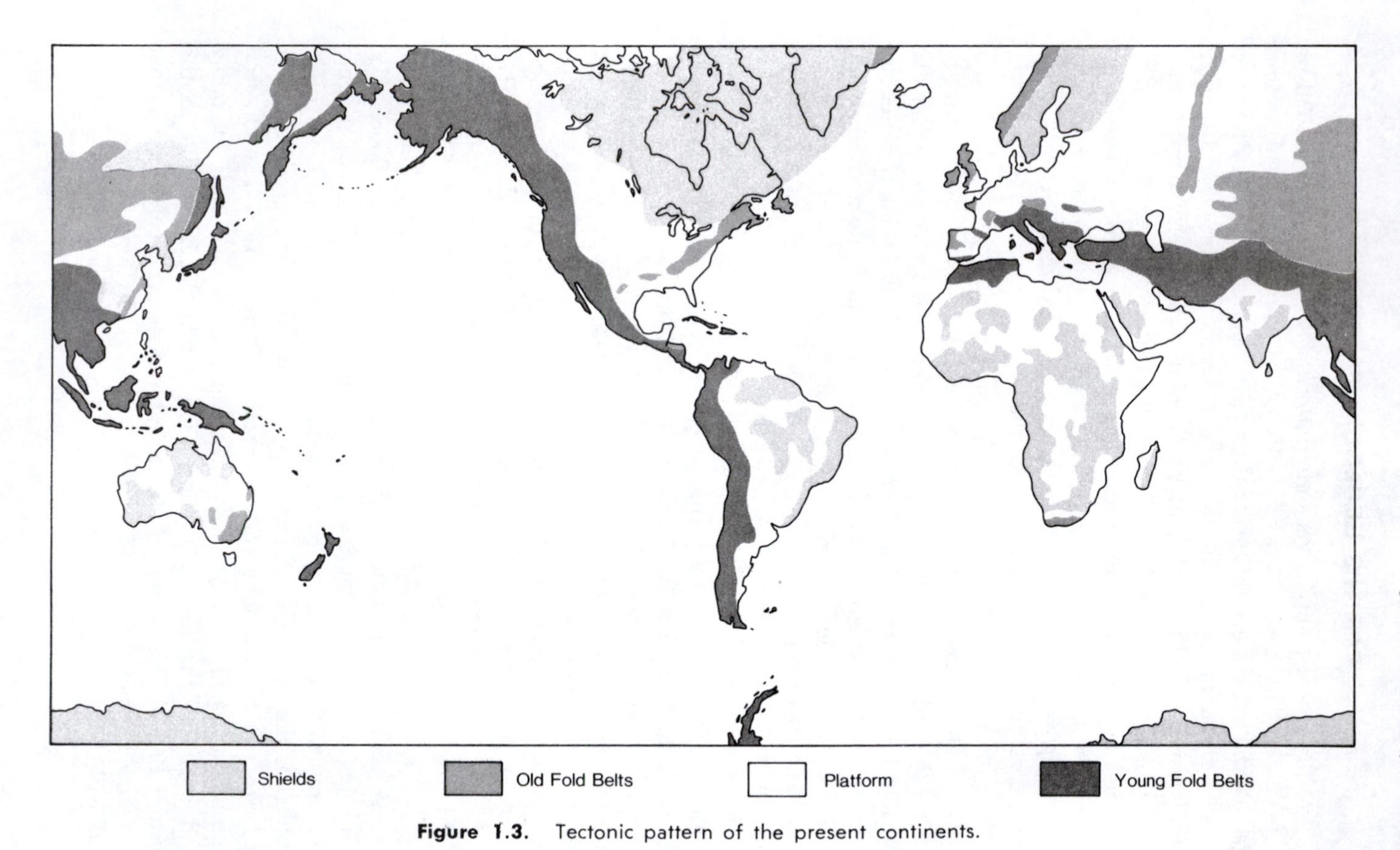

Figure 1.3. Tectonic pattern of the present continents.

with subduction being the most numerous, deepest, and most severe. In the Atlantic Ocean there is no major trench system. The oldest part of the Atlantic floor dates from the time that the Atlantic Ocean came into being when Europe, and later Africa, began to move away from North and South America. The Red Sea of the present day is an incipient ocean, comparable to the early Atlantic Ocean of 200 million years ago.

DEFINING ROCK BIRTHDAYS

In one sense, all earth materials are the same age: they date from the origin of our universe. Yet just as we, individuals made of elements of this same ancient origin, date ourselves as of our date of birth, so do we think of rocks as having a birthday. In the case of sedimentary rocks, the birthdate is the date of deposition. With igneous rocks it is the date they cool. As erosional and tectonic processes go on, rocks are constantly being born, consumed, and reborn again. As bystanders we observe the transitory age relationships of the present. By defining rock birthdays as above we say that some rocks are "older" than others. In a stack of sedimentary strata the layers on the bottom of the stack are the oldest because they were the first to be deposited. This is sometimes referred to as the "law of superposition." Igneous intrusions are always younger than the rocks they invade; faults are always younger than the rocks they cut.

DATING AND CORRELATING ROCKS

Today we have a firm and well-supported basis for assigning ages to rocks. By using rock-dating techniques we can say that certain rocks are so many millions of years old with a fair degree of confidence. The technical ability to do this with confidence has been available for only the past few decades. Before then rocks were assigned ages based on their relative position ("superposition") in the sequence and on the basis of fossils they contained. It was noted empirically that the oldest rocks contain almost no evidence of life, and that as we move upwards in the stratigraphic column fossils appear in some abundance in rocks 600 million years old and younger. In general, the early fossils are less diverse and represent simpler life forms than those that appear later in the fossil record.

Stratigraphic sequences in certain areas were designated as standards for each portion of geologic time as it is represented by rocks. In general, the oldest rocks were represented by standards in the Canadian

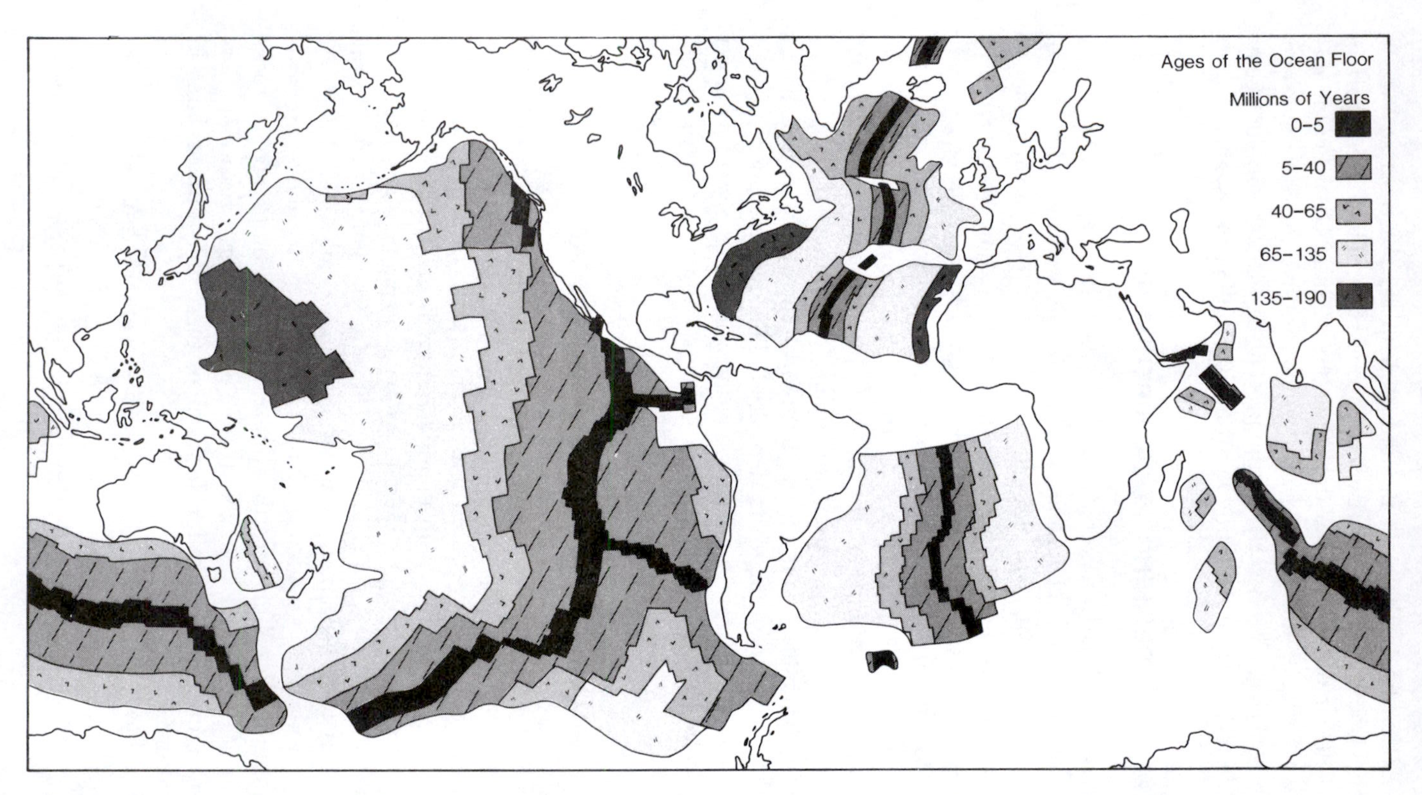

Figure 1.4. Age of origin of the sea floor. Magnetic patterns in sea floor rocks make it possible to extend dates obtained from limited drill samples over most of the rest of the ocean basin.

Shield, intermediate age rocks were represented by rock sequences in Great Britain, and the youngest rocks were represented by standards in central Europe. The nomenclature developed during the last half of the nineteenth century was adopted by geologists throughout the world. It proved its utility by being adapted to show rock ages on geologic maps from any continent and in any language. Geologic time names in common usage are shown on figure 1.5. Anyone using geologic maps or considering geologic history must have this sequence of names well in mind.

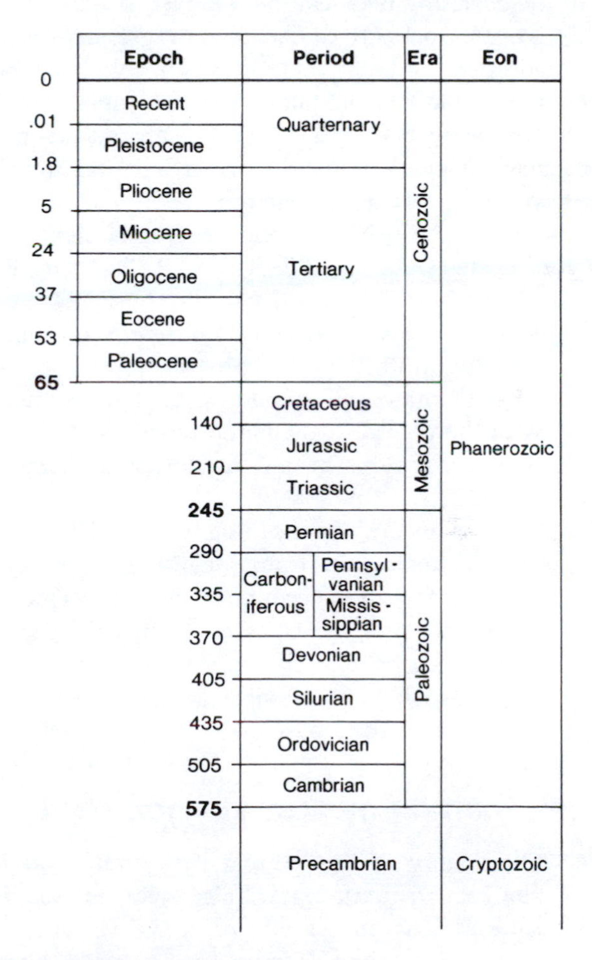

Figure 1.5. Geologic time scale. Numbers show the age in millions of years before the present.

Although the geologic time names on figure 1.5 have been in use for more than a hundred years, the assignment of ages in millions of years has been accomplished only recently after a great deal of data gathering. The best method for determining the actual age (in millions of years) of a rock is by radiometric dating, a process explained in more detail in the next chapter. Radiometric dating works best on igneous rocks. Sedimentary rocks must be dated chiefly by their relationship to igneous rocks. For example, certain stratigraphic sequences may include ashfalls or lava flows. These can be dated directly using radiometric techniques and the age of the enclosing beds can be deduced from their relationship to the adjacent igneous rocks. Radiometric dating is expensive and time consuming, in addition to being chiefly applicable to igneous rocks. Therefore, the age of most sedimentary strata is determined by fossils that have been indirectly tied into the radiometric scheme. Fossils are common, particularly in shallow-water marine deposits. Most animal and plant groups have changed rapidly enough throughout the ages so that their fossil remains can be used to assign rocks sequentially to one of the time categories listed on figure 1.5. It is a bit like using Roman coins to date archeological works.

Earth's magnetic field has reversed itself, north to south and vice versa, frequently enough in its history to provide another basis for dating rocks. Rocks formed at midocean spreading centers, particularly, have been imprinted with the record of magnetic reversals. Magnetic reversal patterns can also be recognized in and used to date some continental rock sequences as well.

Correlation of rocks means determining the equivalence in age or stratigraphic position of rock units from separated areas. A great many schemes have been devised to determine rock equivalences; most are based upon comparison of rock types, fossil occurrences, or radiometric or magnetic dating procedures. Correlation is essential to deducing earth history. We have to know the ages of rocks before we can tell what was going on at any given time in the past.

RECOGNITION OF PAST ENVIRONMENTS

Every rock tells a story. By examining rock strata and interpreting our findings on the basis of uniformitarianism we can conclude, for instance, that New Jersey was once a site of active volcanism, Iowa was often beneath shallow seas, and Illinois was covered by glacial ice. Extrapolating into the past on the basis of fossils and sedimentary or volcanic structures seems reasonable enough. Naturalists have done it through the ages, even before the word "geology" was coined. Paleoenvironmental reconstruction is still a useful mental exercise, not only aesthetically satisfying but often of considerable economic importance. Petroleum geologists search for elusive oil traps by restoring in their mind's eye buried limestone reefs by comparing them to modern reefs forming in the Caribbean; offshort sand bars are visualized beneath Wyoming; stream channel "shoestring" sands are reconstructed beneath Oklahoma.

A more sophisticated line of reasoning enables us to extrapolate plate tectonic relationships into the past. Because we are able to measure present plate movement directions and velocities (directions are shown by arrows on fig. 1.2) we can mentally reverse the direction and transport the plates back to earlier positions. On this basis we can close the Atlantic Ocean by juxtaposing the eastern United States and western Africa 200 million years ago. The velocity-distance plate extrapolation can only carry us back as far as the sea floor magnetic record extends, about 200 million years, as shown on figure 1.4. To deduce plate tectonic relationships existing prior to that we must use other lines of thought.

Although our firmest plate tectonic extrapolations come from the study of the sea floor, a comparatively simple place geologically, recent geologic studies attempt to use plate tectonic concepts to explain some of the more complex geologic relationships we see on continents. To do this we again apply the principle of uniformitarianism to make a deduction. For example, the west coast of South America bears the Andes Mountain chain, which is capped with volcanoes made chiefly of an igneous rock called, appropriately enough, andesite. Just offshore from the Andes, along a major trench system (figure 1.2), the Nazca plate is presently sliding beneath western South America. Geologists have concluded that the andesitic volcanoes are directly related to this kind of plate relationship because similar volcanoes are found almost everywhere this relationship exists today. Therefore, when we find large quantities of 150-million-year-old andesitic rock in central Nevada, we infer that a trench-subduction zone situation must have existed in western Nevada and California at that time and that it produced the andesitic rock.

Of the three major kinds of plate boundaries, spreading centers, subduction zones, and transform faults, only the first two are characterized by particular rock assemblages. Transform faults are marked only by the considerable separation of formerly joined geologic features. As previously mentioned subduction zones are marked by andesitic volcanism on their inner margins. Ancient subduction zones may also be marked by an accumulation of much-mixed open-ocean and near-shore deposits that have been scraped off the top of the descending lithospheric plate.

Occasionally even portions of basaltic sea floor are left behind as the plate descends. The entire mixed-up rock assemblage is known by the descriptive name melange. Paired belts of melange and andesitic rocks doubly identify a subduction zone. Melanges formed in earlier times are excellently exposed in places along the present California coast.

New material formed at midocean spreading centers consists of basaltic pillow lavas, gabbros, and the somewhat denser peridotites. Collectively this spreading center rock assemblage is called ophiolite. Ophiolite suites constitute the lithosphere of the sea floor except for the open ocean and near-shore sediments that rain down on them. Extensive as they are in ocean basins, ophiolite suites are uncommon among continental deposits because they are recycled with the descending plate in subduction zones. Only rarely does a large ophiolite block escape recycling by being emplaced onto the overriding continental plate. The largest such block in North America appears to be a block of Ordovician age located in central Newfoundland.

NORTH AMERICA'S PAST POSITIONS

We have discussed how the magnetic reversal record of the sea floor can be used to trace the relative geographic positions of continents backward about 200 million years. To go back farther in time we must employ another kind of magnetic recording that is found in continental rocks. During the deposition of iron-bearing sediments (or magnetite-bearing ashfalls or lava flows), while the magnetically polarized particles are in transit, they tend to line up with the earth's magnetic field. When these tiny magnets come to rest, they indicate, in aggregate, the position of the earth's magnetic pole at that time. If we check the magnetic polarity of a succession of rocks of different ages from any continent, we find that they show a systematic chronological change in direction through time as though the earth's magnetic pole was moving. Comparison of the records from two continents shows that their paleopolar plots diverge as shown on figure 1.6.

By utilizing paleomagnetic data from all continents, British geologists have reconstructed a picture of the past 500 million years of world geography, as shown on figure 1.7. Lest the maps seem to be the final word, we should note that tracing paleopoles has one important deficiency: we cannot recover the original longitudinal position from rock samples because magnetic inclination for a given latitude is the same for all longitudes. Thus, in tracing continental positions beyond 200 million

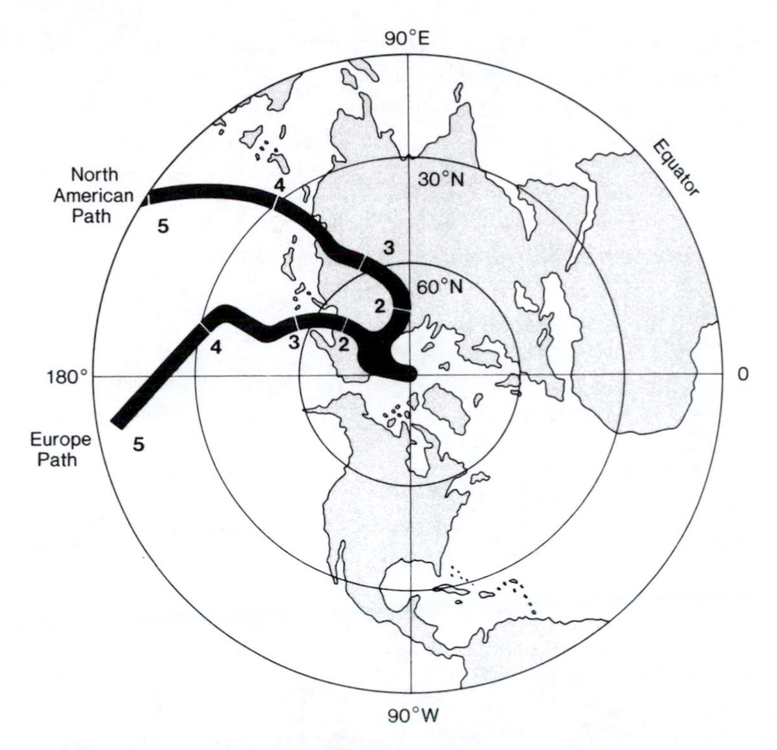

Figure 1.6. Apparent polar-wander paths for North America and Europe. Bold numbers show radiometric dates of rocks used in hundreds of millions of years. Actually the earth's pole did not wander as the figure implies; it remained pretty much in place while the continents moved. The European and North America paths would have been the same if the Atlantic Ocean had not opened about 200 million years ago.

years (the sea floor limit), there is a fair degree of uncertainty about the longitudinal positions of continents.

In summary, paleomagnetic data, both that from the sea floor and that from continental rocks, strongly support the idea of continental mobility. In addition, distribution of certain kinds of rocks and fossils reinforces this concept. Coal found in Antarctica, for example, suggests that the Antarctic continent was in another geographic position when the coal was forming from plant growth. Throughout geologic time, continents have joined and separated, and oceans have opened and closed. Our challenge is to reconstruct the pattern as well as we can.

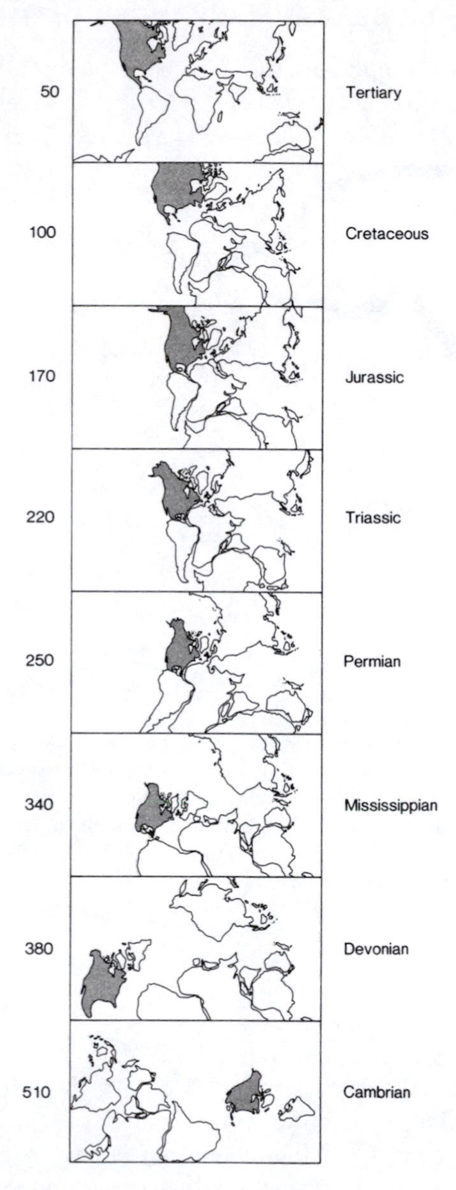

Figure 1.7. A succession of maps showing relative continental positions throughout Phanerozoic time as interpreted by A. G. Smith and J. C. Briden, and G. E. Drewry of Great Britain. Numbers show millions of years before the present. Positions are based primarily on paleomagnetic data. Apparent distortion of the shape of continents is caused by their changing positions on this type of a world projection.

WORLDWIDE SEA-LEVEL CHANGES OF THE PAST

When the geologic periods (fig. 1.5) were being named more than a hundred years ago, their boundaries were selected at unconformities, breaks in the stratigraphic record in the local section selected for the standard. At that time these unconformities were perceived as having worldwide extent, the result of some great catastrophic event. However, studies in this century have shown that this perception was wrong; the unconformities proved to be only of local extent, and many places were found where deposition of strata continued uninterruptedly from period to period. The formally named periods are not the natural geologic packages they were once thought to be.

A vast amount of geologic information has accumulated in the past hundred years, and because oil often occurs near unconformities, petroleum geologists using computers have reviewed this information in their renewed search for worldwide breaks in the stratigraphic record. Using seismic studies on land and sea, radiometric and paleontologic dating procedures, and paleoenvironmental studies, teams of geologists have succeeded in tracing 13 unconformities, shown on figure 1.8, throughout the world. Three of these coincide with period boundaries; the rest do not. Had the geologic founding fathers had access to the present sophisticated information, our time boundaries might have matched worldwide unconformities more precisely!

What could cause a worldwide unconformity? One explanation comes readily to mind: glaciation. When an inordinant amount of the world's water is bound up in a continental icecap, the ocean levels fall. Conversely, as the world's ice melts, the ocean levels rise. By this process worldwide sea-level changes of several hundred feet can be produced in a geologically short time. Phanerozoic world glaciations that could have influenced sea-level changes are known from the Ordovician, the Permian, and the Quaternary.

Another phenomenon that might cause the ocean to spill over the continents would be a bulging of the sea floor. Midocean spreading centers are marked by tremendous mountain chains, such as the Mid-Atlantic ridge system. The rise of hot material along the axis of the spreading center causes the ridge height to be maintained. As material moves away from the axis and cools, it subsides in elevation ultimately to the level of the abyssal sea floor. If hot material were added at the spreading center at an accelerated rate, the midocean ridge would become broader and take up more of the ocean basin. But if new material were added more slowly at the axis, the flanks of the ridge would decline and the volume of the ocean basin would increase. This would cause a

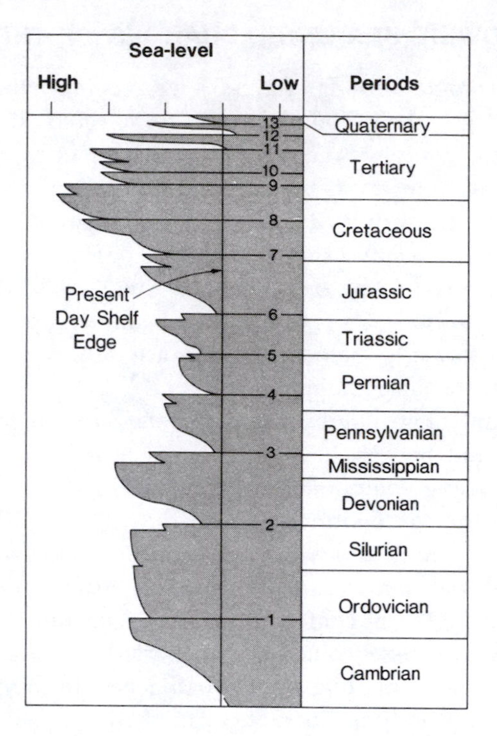

Figure 1.8. Major global unconformities. Major advances of seas over continents are indicated where the shaded area extends to the left. Rapid sea withdrawal into ocean basins is indicated by horizontal lines. (After P. R. Vail and R. M. Mitchum, Exxon Corporation, 1978)

withdrawal of oceans from the continental margins and bare the continental shelves to erosion producing a worldwide unconformity.

The pattern of sea-level changes shown on figure 1.8 indicates that the advances were slow protracted affairs, while the withdrawals happened quite rapidly. There is some evidence to suggest that global unconformities may coincide in time with major mountain-making events (orogenies). Since orogenic activity is probably controlled by changes in plate tectonic forces, the coincidence supports the view that unconformities are also related to changes in plate motion. Unfortunately, the fundamental cause of change in plate tectonic activity is not known.

2

Precambrian

TOPICS

FORMATION OF THE EARTH

The planet earth came into being 4.6 billion years ago when a great cloud of dust and gas collapsed and condensed to form our solar system. Earth's primordial nucleus developed at such a fortuitous location within the nebular disk of the forming solar system that it attracted a great proportion of heavy elements and became the densest planet in the system. However, because of its small size and nearness to the proto-sun, the earth lost a disproportionate share of lighter elements and compounds. Heat from short-lived radioactive isotopes in the cloud, added to heat from gravitational infall and meteorite impact, caused the incipient earth to partially melt and undergo differentiation into a fluid core surrounded by a mantle and outer crust. During its initial hot stage much of earth's nickel and iron sank to form the earth's central core, while oxides and silicate minerals rose and differentiated to form the mantle and outer crust. As the lighter elements, such as silicon, aluminum, sodium, and potassium, rose toward the top, they were accompanied by certain heavy radioactive elements, uranium and thorium, whose chemical affinities caused them to be concentrated with the lighter crustal-forming elements despite their weight.

The foregoing scenario is an explanation for the observed major layering of the earth. It also accounts for the observation that the earth's present internal heat is being generated chiefly within the crust and upper mantle by the following long-lived radioactive isotopes that are concentrated there: thorium 232, uranium 238, potassium 40, and uranium 235.

No terrestrial rocks retain a radiometric age older than 3.8 billion years. However meteorites, believed to represent relict material from the early solar system, were solidified 4.6 billion years ago according to their radiometric dates. Even though no 4.6 billion-year-old rocks have remained unaltered on the earth, we can deduce the age of the earth itself by tracing the change through the time of the relative abundance of lead 206 and lead 207, inert daughter products respectively of radioactive uranium 238 and uranium 235. Extrapolated backwards, the lead 206/207 ratio has a pattern that indicates that 4.6 billion years ago the isotopic abundance ratios of lead were identical in the parent materials of both meteorites and the earth. This is one of the bases for concluding that meteorites, earth, and the rest of the solar system (see fig. 2.1) had a common time of origin 4.6 billion years ago.

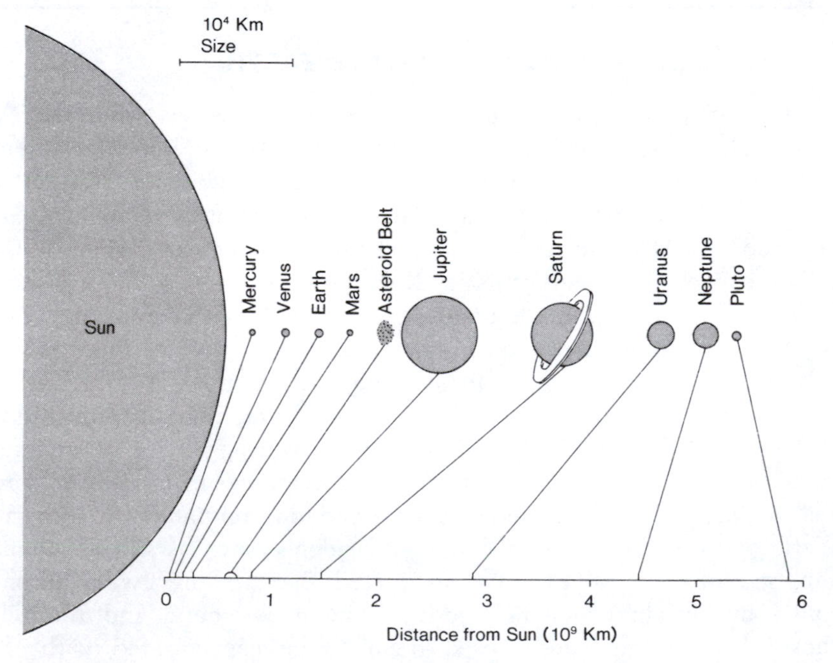

Figure 2.1. Our solar system, showing the relative size and average distance from the sun to the planets. The spacing, distribution in a disk, and consistency in direction of rotation of the planets led observers long ago to the view that the members of the solar system share a common origin.

Careful and thorough searching on all continents by geologists seeking the earth's oldest rocks has not revealed any that are older than 3.8 billion years. Sophisticated radiometric dating procedures that can see through minor date-modifying heat episodes have made it possible to identify rocks belonging to the 3.0-3.8-billion-year-old time span on most continents. It now seems unlikely that we will ever find rocks formed on earth that have retained radiometric dates older than 3.8 billion years. Events in the first 0.8 billion years of earth history must have involved so much heat that no earthly rock body could remain cool enough to preserve intact the radioactive parent-daughter relationship necessary to date rocks.

Although rocks retaining radiometric dates older than three billion years are very scarce here on earth, they make up almost all of the moon's rocks. We can, therefore, turn to the moon for a better record of the early history of the solar system.

HISTORY OF THE MOON

Man's conquest of the moon in the last half of this century was not only a technological triumph but also a breakthrough in our knowledge of early events in the history of the solar system. The moon was studied by the most sophisticated tools available: mass spectrometers, nuclear reactors, remote X-ray sensors, microscopes, magnetometers, and seismometers. And from this intensive study we have put together a coherent outline of the moon's history that, by analogy, tells us about obscure aspects of the earth's earliest history. Our present understanding of the moon's development can perhaps be best summarized by discussing the sequence of stages shown in figure 2.2.

I. *Planetesimal accretion.* The moon and the earth most likely began their development together in the primordial solar nebula with the moon early on acquiring its role as a satellite to the earth while both were still growing by accretion of planetesimals. Planetesimals are small grains or clusters that condensed out of the hot nebular gas, those minerals with the highest vaporization temperatures being the first to form. The accreting moon seems to have captured a greater proportion of high-temperature condensate particles than the nearby earth; we conclude this because moon rocks are systematically higher in high-temperature element, such as titanium, and lower in more volatile elements, such as sodium, potassium, and bismuth, than are earth rocks.

It has been estimated theoretically that the moon could form by accretion from a disk of planetesimals in as little as 1000 years, but we

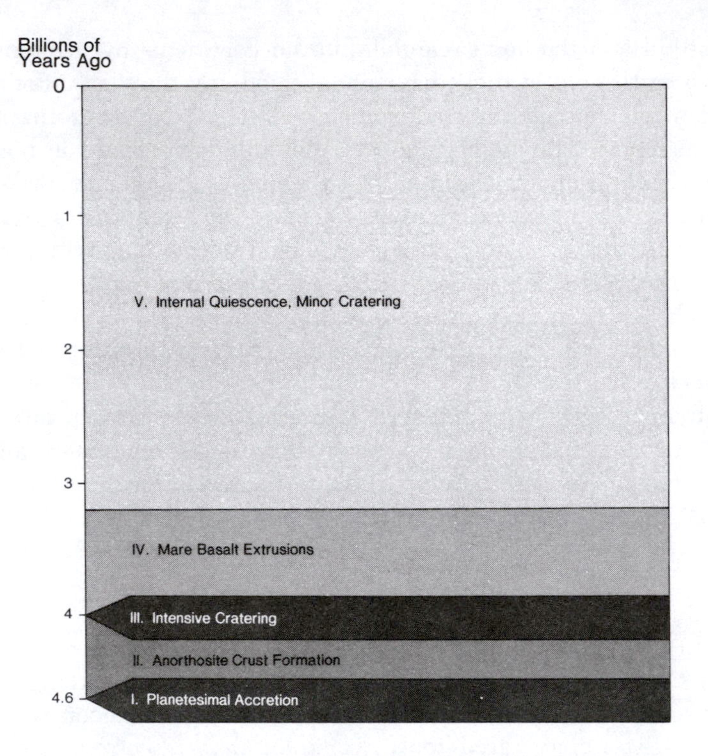

Figure 2.2 Major stages in the history of the moon plotted on a radiometric time scale. Note that all of the moon's internal activity occurred more than three billion years ago and that, except for some cratering of its surface by meteorite infall it has been dead since then.

really have no way of measuring this. However, from its relation to later lunar events we can surely say that the moon's birth took an astronomically short time.

II. *Anorthosite crust formation.* Although Apollo astronauts collected as great a variety of rocks as they could from six points on the moon that had been carefully selected for their heterogeniety, nearly all samples obtained can be classified as one of two kinds of rock: mare basalt and anorthosite. Anorthosite, an igneous rock composed almost entirely of calcium-bearing plagioclase feldspar, is overwhelmingly the most abundant type of rock on the moon. It makes up almost all of the lunar highlands. That is to say, it is found everywhere except for those areas on the near side of the moon that are covered by dark mare basalts. All lunar anorthosite has been brecciated by intensive bombardment by meteoroids after it formed. Heat produced by the bombardment reset all radiometric clocks in the anorthosite to about 4.0 billion years, even though the anorthosite had solidified prior to that. Much of the material collected by the astronauts had its origin tens of kilometres beneath the moon's surface, but was brought to the surface by huge crater-forming impacts. It is believed that the anorthosite makes up a crustal layer on the moon about 60 km thick. Seismic velocities (speed of moon-quake waves) above a depth of 60 km are consistent with those for anorthosite; velocities measured below 60 km indicate that a rock of higher density makes up the lunar mantle.

III. *Intensive cratering.* Both the crater-pocked surface of the old anorthosite highlands and the brecciation and pulverization of all highland rocks testify to the relentless bombardment of the moon's surface in its early history. The oldest radiometric dates from the moon, those of the highland rocks, tend to cluster around 4.0 billion years, although the highland anorthosite actually crystallized earlier than that.

It would appear that some violent heat-producing event reset all lunar radiometric clocks about four billion years ago. This event was probably the colossal bombardment that created the Imbrian and other large basins on the near side of the moon. Debris from these impacts blankets much of the moon's near side. Possibly a late-arriving group of planetesimals was encountered about four billion years ago, resulting in a great surge of cratering on the moon and other planets. Craters of this antiquity cannot be identified on the earth because the mobility of the earth's surface has long since destroyed any physiographic expression of such features. However, the fact that we cannot find any earthly rocks with radiometric ages much older than 3.8 billion years suggests that all radiometric clocks on earth may have been reset by heat-generating meteoroid bombardment during the earth's early history. Our solar system must have been an inhospitable place four billion years ago!

IV. *Mare basalt extrusions.* Galileo, more than three hundred years ago, gave the name *maria* (Latin: seas; singular—*mare*) to the dark areas on the moon's near side. We now know that these are basalt floods that filled basins previously excavated by meteoroid impacts. Mare lavas issued intermittently over a period of about half a billion years (3.8-3.26 billion year ago) from fissures rooted in the moon's upper mantle. They form a sequence of flows whose relative ages can be readily deduced by their overlapping relationships. Radiometric dates tell us that some of the most extensive flows in the Imbrian Basin solidified 3.3 billion years ago.

Composition of mare basalt varies with its age, the oldest samples having the highest titanium content. This suggests that the source of the basalt deepened in the mantle with time, as the radioactivity peak, the heat source, migrated inwards. With such extensive lava fields as the lunar maria, we might expect to find numerous volcanic cones and caldera, but such is not the case. Only a few low domes and cones, similar to small shield volcanoes on earth, have been identified as lunar volcanoes.

By a little over three billion years ago the moon's crust and outer mantle had cooled to the point that the moon had developed an armor too thick to permit any further penetration of basalt to the surface. The moon's igneous activity had come to an end.

V. *Internal quiescence, minor cratering.* Maria surfaces are almost unscarred by large craters, showing that by the time of their formation major impacts had ceased to be common. Close up, they are seen to be pocked with small craters, the continuing effect of bombardment by small objects. Earth's atmosphere has shielded its surface from most small meteoroids. The few large impact craters that have formed in the last stage of lunar history are identified by rays, bright streaks of glassy material radiating outward from such craters. Superposition and cross-cutting relationships permit us to determine the relative ages of younger impacts.

Seismic instruments placed on the moon by the astronauts tell us that the lunar lithosphere at the present time constitutes an immobile armored layer about 1000 km thick. There is no plate tectonic activity possible on the moon under these conditions. This contrasts with our earth, whose thin lithosphere is in constant motion. Except for meteoroid bombardment, the moon's surface has been inactive for the past three billion years, a period of time during which the Cryptozoic and Phanerozoic rock record here on earth was being made.

EARTH'S EARLY ATMOSPHERE, OCEANS, AND CRUST

The observation that the inert gases, neon, krypton, and xenon, are a billion times less abundant on earth as compared to the rest of the universe has been interpreted to mean that the earth was swept clean of its earliest atmosphere by solar winds. If this is so, the earth's present atmosphere must have developed by release of gases from its interior. Volcanoes continue this outgassing today by discharging water vapor, carbon dioxide, sulphur dioxide, nitrogen, hydrogen, argon, and chlorine. It has been suggested that early emanations probably included methane and ammonia, both compounds derivable from nebular material, such as the carbonaceous chondrite meteoroids that may have been included among the planetesimals that formed the earth. Methane and ammonia are the major components of Jupiter's atmosphere today.

Free oxygen is not emitted by modern volcanoes, nor was it in the past. Certain minerals from the oldest rocks indicate that the amount of free oxygen in the primeval atmosphere was extremely small. Some oxygen developed in the atmosphere, as shown in figure 2.3, from the dissociation of oxides, such as carbon dioxide and water. This process occurs slowly in the upper atmosphere using the energy of ultraviolet light. It also occurs much more rapidly through the process of organic photosynthesis. Chlorophyll-producing microorganisms have been identified in South African rocks three billion years old; but the process of photosynthesis may have started 3.7 billion years ago. Evidence for this is

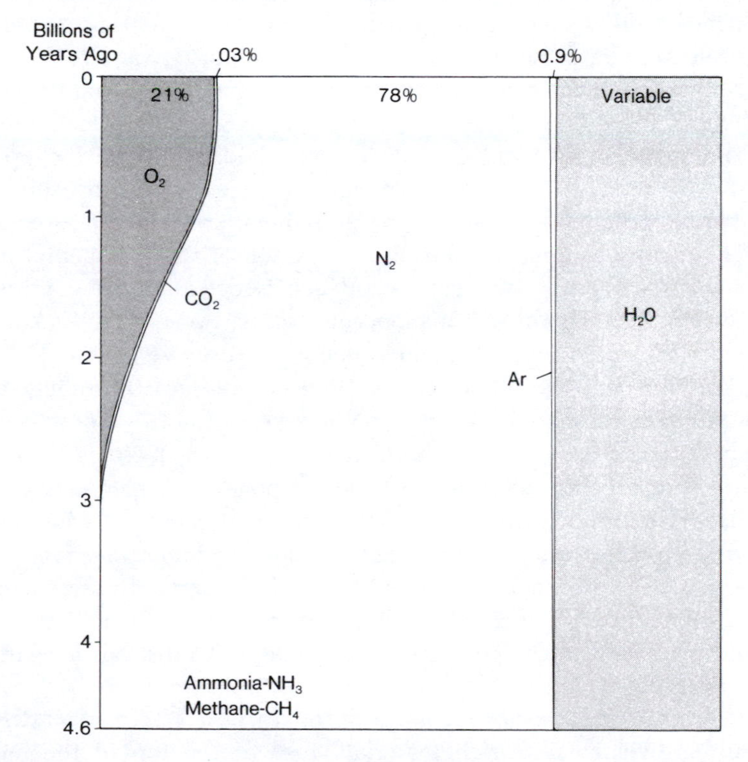

Figure 2.3. Change in composition of earth's atmosphere with time. Early methane and ammonia have changed to water, carbon dioxide (or carbonate in rocks), and nitrogen. Free oxygen has been liberated from carbon dioxide by photosynthesis ever since the appearance of chlorophyll-bearing plants. Present atmospheric percentages are shown along the top line.

indirect, based upon the idea that iron acted as the first acceptor of free oxygen produced by photosynthesizing bacteria and algae. Banded iron formations in West Greenland, dated at 3.7 billion years, may have used the first free oxygen available to form the iron oxide mineral, hematite. Later build-up of appreciable free oxygen in the atmosphere paralleled the proliferation of blue-green algae in the Precambrian seas. By about two billion years ago sufficient oxygen was being produced to not only satisfy mineral oxidation needs but also to produce an oxygenic atmosphere.

Astronauts remarked that the blue color imparted to earth by its oceans makes it the most inviting body in the heavens. We tend to think of the oceans as having always existed, but during earliest earth history, when planetesimals were raining down, radioactivity was intense, and solar winds were sweeping the earth, it is unlikely that oceans could have been present. However, by about four billion years ago temperatures had fallen enough so that water could condense. Some of earth's oldest rocks from West Greenland, dated at 3.8 billion years, include lavas and sedimentary rocks that were water-laid. Although the present quantity of ocean water seems vast, it can be explained by the same process that produced the atmosphere, namely, release of gases from rocks of the crust and mantle during the growth of the continental crust.

About 30 percent of the earth's surface is occupied by continental crust, which is much thicker than oceanic crust, averaging 40 km and ranging to 70 km in mountain roots; oceanic crust averages only 6 km thick. Continental crust is much older than oceanic crust, ranging to 3.8 billion years as compared with 200 million years for the oldest oceanic crust, because oceanic crust is "born again" by plate tectonic recycling. Continental crust and oceanic crust float upon the upper mantle; the three have average densities of 2.7, 3.0, and 3.4 respectively. Most of the upper continental crust consists of igneous and metamorphic rocks with an average composition of granodiorite, and veneered by sedimentary rocks. In mountainous areas continental crust has complicated structure and variable chemical composition, whereas oceanic crust has a relatively simple layered structure and uniform composition.

Continental crust began forming in the earliest stages of earth history, as lightweight silica-rich materials rose to the top of the mobile mantle. The volume of the earth's upper mantle is two-thirds that of the whole earth, a volume large enough to have produced all of the continental crust (only 1 percent of the earth's volume) through geologic time without having undergone appreciable change in composition itself. Continental crust which formed as early as 3.8 billion years ago has been identified on several continents; the extent of continental crust at this early time was between 10 and 40 percent of the present continental area. Irreversible chemical differentiation of the upper mantle throughout time has resulted in the predominance of continental growth over continental erosion and recycling. In fact, continental rocks are so lightweight that they are hardly recycled at all by plate tectonics. Continents today are higher, grander, and more extensive than they have ever been in the geologic past! On the other hand, old oceanic crust is in scarce supply; only those fragments of oceanic crust that have clung to continental edges have escaped recycling.

RADIOMETRIC DATING

The first estimates of the age of the earth were made by adding up the life spans of people mentioned in the Bible. Beginning in the 1700s scientists began trying to use other rate processes, such as the rate of cooling of the earth from an assumed original molten state, rate of accumulation of salt in the seas, rates of accumulation of sediments, and rates of erosion. Each of these ways of estimating earth ages gave values, surprising in their day, of hundreds of thousands to millions of years. But even geologists were surprised when, at the turn of this century, radiometric dating gave the first rational estimate of the age of the earth and also of the great length of Precambrian time. Physicists told the world that our oldest rocks must be *billions* of years old rather than the few millions of years previously estimated.

Modern age assignments of many geologic phenomena are dependent on a group of laboratory techniques known collectively as radiometric dating. Decay of radioactive material occurs at a regular rate, the rate being characteristic of each particular elemental material, or isotope. We express this decay rate as the "half-life" of the isotope, meaning the time it takes for half of any given amount of parent isotope to transform into its daughter products, isotopes of other elements. Although halving something means you always have half of it remaining, in actual fact there is a limit to the detectability of a substance, and a point is reached where, for all practical purposes, the material is gone. Figure 2.4 shows that after 10 half-lives have elapsed, less than one-thousandth of the original amount remains. Similarly, after 20 half lives, less than one-millionth would remain; after 30 half-lives, less than one-billionth. If a highly radioactive isotope produced by an atomic blast had a half life of exactly one minute, this would mean that 30 minutes (half lives) after it came into being only one-billionth of it would remain. So it is that,

Fraction left 1	½	¼	⅛	1/16	1/32	1/64	1/128	1/256	1/512	1/1024
Elapsed half-lives	1	2	3	4	5	6	7	8	9	10

Figure 2.4. Relation between number of elapsed half lives and amount of radioactive isotope remaining. After ten half lives less than one-thousandth of the original amount is left.

except for radiocarbon, which is constantly being produced by cosmic radiation in the upper atmosphere, only those radioactive isotopes with extremely long half lives have endured from their nuclear birth in the formation of the solar system until the present day.

Some 90 chemical elements make up earth's rocks; each element is represented by from 1 to 10 naturally-occurring isotopes, so that the total number of natural isotopes is about 300. Of these 300, two dozen are naturally radioactive. Lead 204 has the distinction of being the natural radioistope with the longest half life: 10^{17} years. Uranium 235 is the primordial isotope with the shortest half life—720 million years.

All radioisotopes could serve as atomic clocks because they all follow the same patterns of decay except at different rates. But experience has shown that certain radioisotopes are much better adapted to rock-dating than others for two primary reasons:

1. They are common in rock-forming minerals.
2. They have half lives less than 100 billion years and hence can have generated detectable amounts of daughter products within the time span of earth history.

The isotopes that are commonly used for rock dating are shown in figure 2.5.

When the half life of an isotope is known, its "age" can be calculated by determining the ratio of parent/daughter isotopes in the sample. Age, in this instance, means the amount of time the daughter isotope has been accumulating around the parent without escaping. Daughter isotopes are always chemically different from their parent. At ordinary surficial temperatures, daughter products remain with the parent. If the rock is heated the daughter isotope tends to migrate; if the rock is melted the parent-daughter combinations are separated completely and the rock's clock is reset at zero when new mineral crystals form. Hence radiometric dates are records of heating events, rather than ultimate age. The remarkable thing is that some rocks have existed on this turbulent earth of ours for more than three billion years without having their clocks completely reset!

Radio Isotope	Half-life in Years	End Product
Uranium—235	720 million	Lead—207
Potassium—40	1.3 billion	Argon—40
Uranium—238	4.5 billion	Lead—206
Thorium-232	14 billion	Lead—208
Rubidium-87	47 billion	Strontium—87

Figure 2.5. Chief decay series of primordial isotopes used for rock dating.

Only miniscule amounts of a radioactive substance are necessary for dating. Except for potassium, the radioisotopes ordinarily occur as a minor impurity in ordinary rock-forming minerals. As the parent isotope decays, the daughter products are retained within the crystal latticework of the host mineral. Minerals differ in their ability to retain various daughter products, and daughter isotopes differ in their escape capabilities. Argon gas, for example, escapes more readily than lead. Hence potassium-argon dates are commonly younger than uranium-lead dates for the same rock. There is, in fact, a "pecking order" in accuracy of radiometric dating procedures. If several kinds of radiometric procedures are applied to the same rock sample, we find that the uranium-lead determination on the mineral zircon will ordinarily yield the greatest age; rubidium-strontium determination on a whole-rock sample will yield a slightly younger age, and potassium-argon determinations will produce successively younger age dates respectively for the minerals hornblende, plagioclase, biotite, and orthoclase. It is evident that certain minerals are better jails for daughter argon than are others.

Fission track dating uses a somewhat different measurement than the parent/daughter ratio of other dating procedures. Fission tracks are damage trails left in a mineral's lattice as an atom of uranium, included within the mineral as an impurity, splits into two middle-sized atoms, which are propelled violently away from their point of origin. The fission tracks, when enlarged by chemical etching, can be counted under a microscope. Their abundance depends on the amount of uranium present and the length of time it has been there. The amount of uranium present can readily be measured so that the number of tracks can be converted to a measure of the age of the specimen. Fission track procedures can be used on a wide variety of materials of a great range of ages. The chief drawback to the method is that minerals differ considerably in their track-healing properties when heat is applied. Tracks in some substances are healed by low temperatures; tracks in all minerals are lost when they are heated above 350 degrees C, well below the melting point of most minerals.

Because the natural radioisotopes listed on figure 2.5 decay so slowly, they are not very suitable tools for measuring recent geological events. Trying to use them for Pleistocene time is a bit like trying to use a yardstick to measure a fly's eye. Other techniques that are applicable to the latest part of earth history are indicated on figure 2.6. Carbon 14 is the most widely used of the younger dating systems. Carbon 14 is not a primordial radioisotope. Rather it is being produced constantly in the earth's upper atmosphere from the transformation of Nitrogen 14 by cosmic ray bombardment. Carbon 14 has a half life of only 5700 years, so that when some of it is removed from the atmosphere by being taken into growing plant (and subsequently animal) tissue, it commences to revert to nitrogen at a steady rate. A bit of wood that grew 5700 years ago is now only half as radioactive as plant tissue growing today. Figure 2.6 shows that each dating system has its best range of applicability.

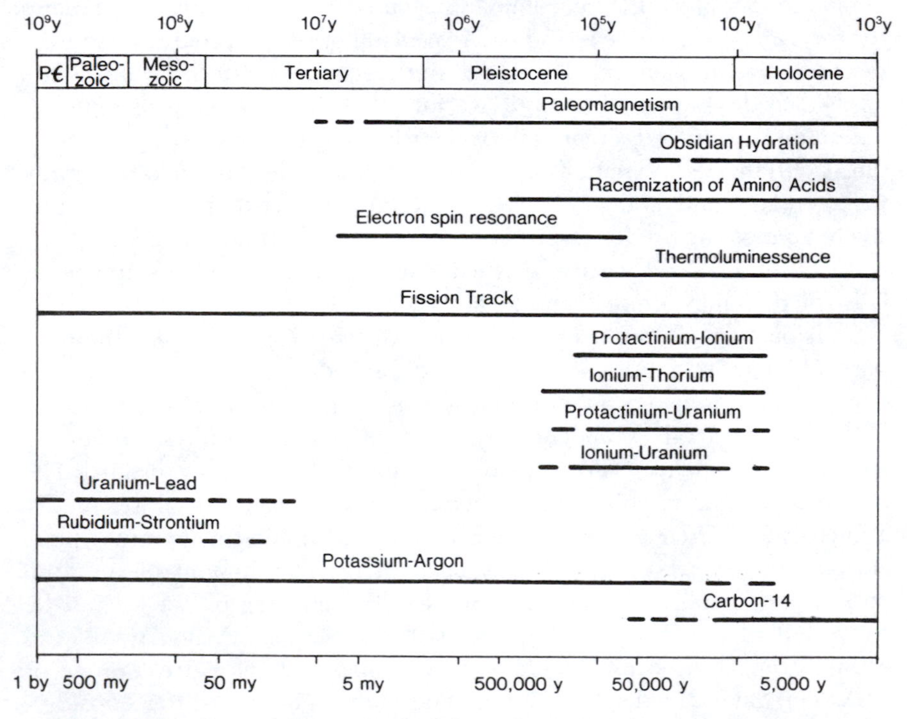

Figure 2.6. Range of age applicability of various dating methods. Note that the time scale is logarithmically shown.

PRECAMBRIAN TIME SCALE

Before the advent of widespread radiometric dating we had no adequate way to classify Precambrian rocks as to age of origin and to correlate Precambrian rocks from continent to continent. For Cambrian and later strata we rely largely on fossils for intercontinental correlation. But fossil remains are sparse in Precambrian rocks and are not generally age-diagnostic. Because of this, various indirect methods were formerly used in deciphering the Precambrian record. The earth was assumed to have evolved from a molten and cooling state when granites and high-temperature metamorphic rocks were formed ("Azoic," "Archean," or "Archeozoic" time) to a more normal condition when familiar sedimentary and volcanic processes prevailed in later Precambrian ("Algonkian" or "Proterozoic") time.

Radiometric dating soon pointed out the fallacy of believing that all Precambrian crystalline rocks (gneisses, granites, etc.) were of "Archean" age. In fact, some highly metamorphosed rock sequences were found to have been formed late in Precambrian time. Even some highly metamorphosed Phanerozoic rocks were originally mistakenly classified as Archean.

Canadian geologists have made thousands of radiometric age measurements of Precambrian rocks of the Canadian Shield (fig. 2.7) and have identified three peak times of radiometric clock resetting. The oldest and most widespread occurred 2500 million years ago and is termed

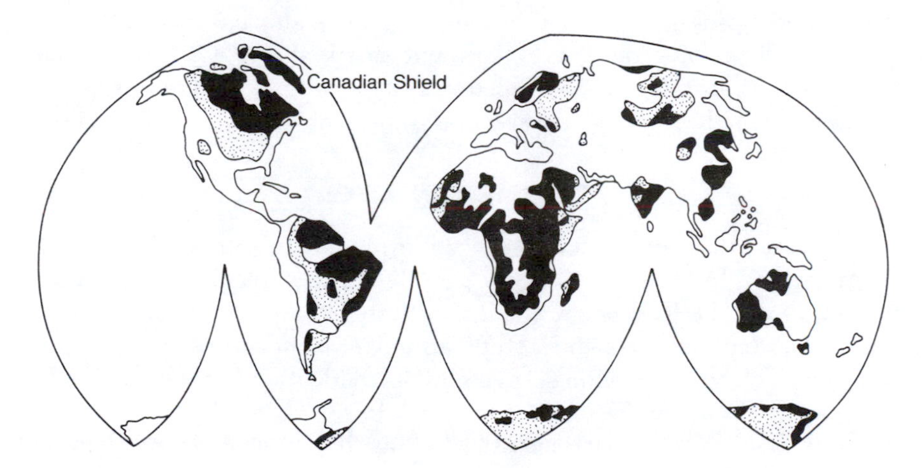

Figure 2.7. Precambrian shield exposures shown in black, thinly veneered shield areas in dotted pattern, areas of thick younger cover in white.

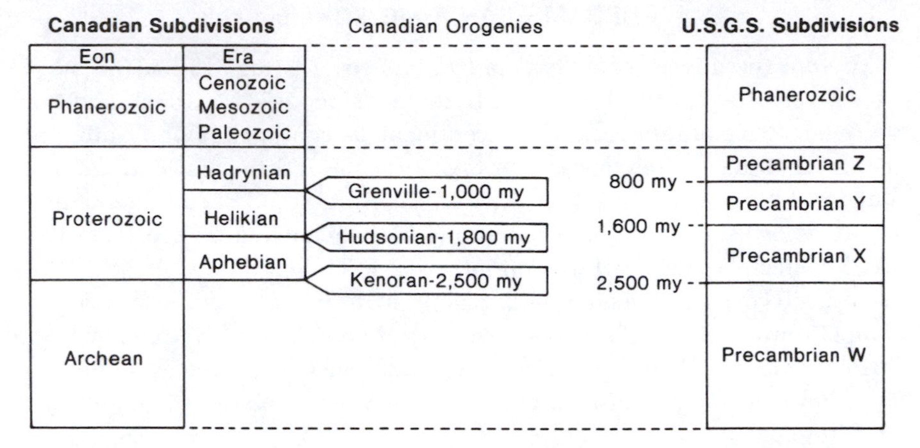

Figure 2.8. Comparison of Precambrian subdivisions used in Canada and the United States.

the Kenoran Orogeny (mountain making). The next, and somewhat less widespread event peaked about 1800 million years ago and is called the Hudsonian Orogeny. The last Precambrian orogeny, the Grenville Orogeny, affected only easternmost North America and peaked about 1000 million years ago. Canadians have used the orogenic peaks to mark the boundaries of their newly established Precambrian subdivisions, shown on figure 2.8. They have retained the old names Archean and Proterozoic but have redefined them as radiometric time intervals, rather than continuing the old usage discussed in the paragraphs above.

The U.S. Geological Survey has not adopted the Canadian time names, but is using a W-X-Y-Z letter system with boundaries slightly differing from the Canadian system, as shown on figure 2.8.

PRECAMBRIAN LIFE

The last few decades have seen an explosion in our knowledge of Precambrian life. A hundred years ago the earliest fossils known were trilobites from the base of the Cambrian System, and one of the greatest puzzles of that day was the abrupt abundant appearance in the rock record of such complex animals. We now have identified numerous kinds of organisms from Precambrian rocks and have extended the record of life back to 3.3 billion years, nearly as old as the oldest rocks recognized on earth.

Perhaps the most fundamental division in living organisms is not that between plants and animals but rather that between organisms whose cells lack nuclei, the prokaryotes, and those whose cells contain nuclei,

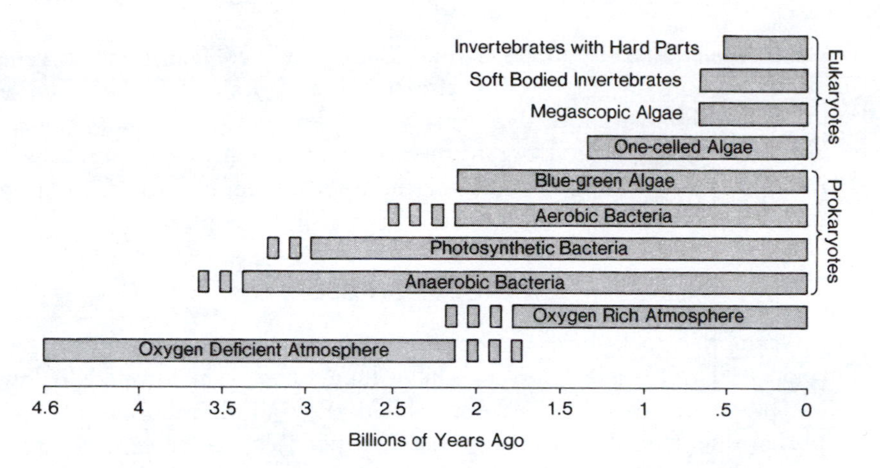

Figure 2.9. Major events in the appearance and progression of life on earth. Simple cells without nuclei (prokaryotes) appeared more than three billion years ago when the earth was little more than a billion years old. The profusion of animal groups familiar today appeared only in the last half billion years.

the eukaryotes. Prokaryotes include only two groups of organisms, the bacteria and the blue-green algae. Bacteria are the oldest known organisms, as shown on figure 2.9.

The progression of life is tied closely to changes in the atmosphere. Earth's primordial atmosphere was deficient in free oxygen and the earliest bacteria lived without it by using fermentation to obtain energy from organic molecules that had formed nonbiologically in the primordial atmosphere and ocean. Photosynthetic anaerobic bacteria appeared about three billion years ago and were followed by aerobic photosynthesizers about a half billion years later.

Oxygen generated by photosynthesizing prokaryotic bacteria and blue-green algae was first taken up by iron dissolved in the oceans, which was then precipitated to form the unique mid-Precambrian banded iron formations. When the oceans had been swept of iron, free oxygen began to accumulate in the atmosphere. At that time anaerobic organisms moved to oxygen-free habitats and aerobic blue-green algae flourished in the shallow ocean waters. Some blue-green algae had the ability to precipitate calcium carbonate and form stromatolites, which are massive reef-like structures. When an oxygen-rich atmosphere was established, organisms evolved that could not only tolerate oxygen but also employed it in respiration, a great advance in metabolic efficiency. About 1.5 billion years ago the first eukaryotic (cells with nuclei) organisms appeared. Diversification of the eukaryotes and the development of sexual reproduction occurred a little more than a billion years ago. The result is seen in Late Precambrian rocks of Australia, Great Britain, and Newfoundland, where the fossil record reveals a complex and diverse assemblage of multicelled, soft-bodied organisms, some of which are as much as a foot long. Thus the stage was set for the appearance, in early Cambrian time, of more complex animals with preservable hard parts.

CANADIAN SHIELD

Precambrian rocks form the nucleus around which each continent has developed. These ancient terranes have long since been beveled to low relief, hence the descriptive name, shield. Present distribution of Precambrian shields is shown on figure 2.7; a possible earlier configuration is represented on figure 2.10.

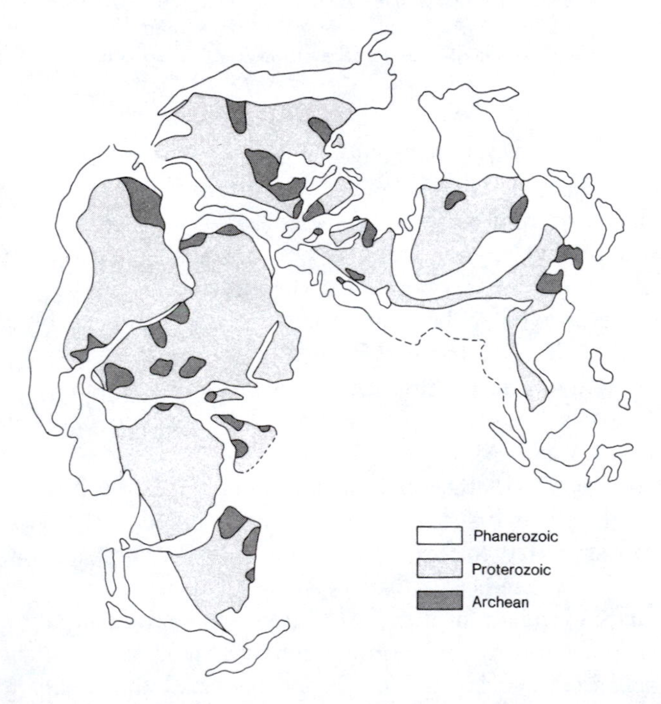

Figure 2.10. Precambrian shields, as they may have been arranged to form a late Paleozoic supercontinent 300 million years ago. A similar clustering of continents is believed to have existed during Precambrian time. Each continent bears old Archean regions surrounded by Proterozoic terranes that are adjoined by Phanerozoic mobile belts. (After Windley, The Evolving Continents, 1977, John Wiley & Sons.)

Archean terranes include two contrasting rock assemblages: highly metamorphosed granitoid gneisses and less metamorphosed greenstones. Some Archean terranes are so metamorphosed that identification of the mode of origin of their rocks is approached chiefly through their mineralogic makeup. Some appear to represent metasediments, including a few marbles, quartzites, and iron formations, while others appear to be metavolcanic and metagranitic rocks. These rocks may have formed in an ancient plate-tectonic arc-trench environment.

Archean greenstones occur in belts that are remnants of ancient volcano-sedimentary basins. Their earliest deposits are predominantly basaltic; these grade upwards into sedimentary strata and andesitic lavas. In plate-tectonic terms they may represent submarine deposition in a basin marginal to the arc-trench environment mentioned above. Archean history of the Canadian Shield culminated with widespread intrusions of the granitoid rocks during the Kenoran Orogeny 2.5 billion years ago (fig. 2.8).

Proterozoic rocks in the Canadian Shield are quite diverse and range from rocks metamorphosed during the Hudsonian and Grenville Orogenies (fig. 2.8) to unmetamorphosed late Precambrian sedimentary and volcanic sequences that retain all of their primary depositional features. Many Proterozoic sedimentary and volcanic sequences are thousands of metres thick, representing great accumulations in local basins. These sedimentary and volcanic rocks are now much less areally extensive than the granites and gneisses with which they are surrounded.

Where rocks of the shield extend beneath younger rocks, the age of the Precambrian basement has been ascertained from radiometric dating of oil well cuttings. A generalized map summarizing dates of Precambrian rocks is given on figure 2.11. Oldest dates are clustered in three areas on the map, areas wherein events younger than 2.5 billion years have not been associated with enough heat to reset the radiometric clocks. Later dates are found on the periphery of the older centers, giving rise to the idea that continental masses grow by accretion of younger materials around their edges.

Paleomagnetic poles plotted for several places in the central and northern Canadian Shield follow a single wander path from 2200 to 1000 million years ago, as shown on figure 2.12, suggesting that the bulk of the shield moved as a single entity during that time interval. Poles plotted from Grenville rocks along the eastern edge of the shield do not fit the main polar wander path. The oldest recognized Grenville path, at 1150 million years, follows a different course than the main part of the shield, but joins up with the main path at 1000 million years from which

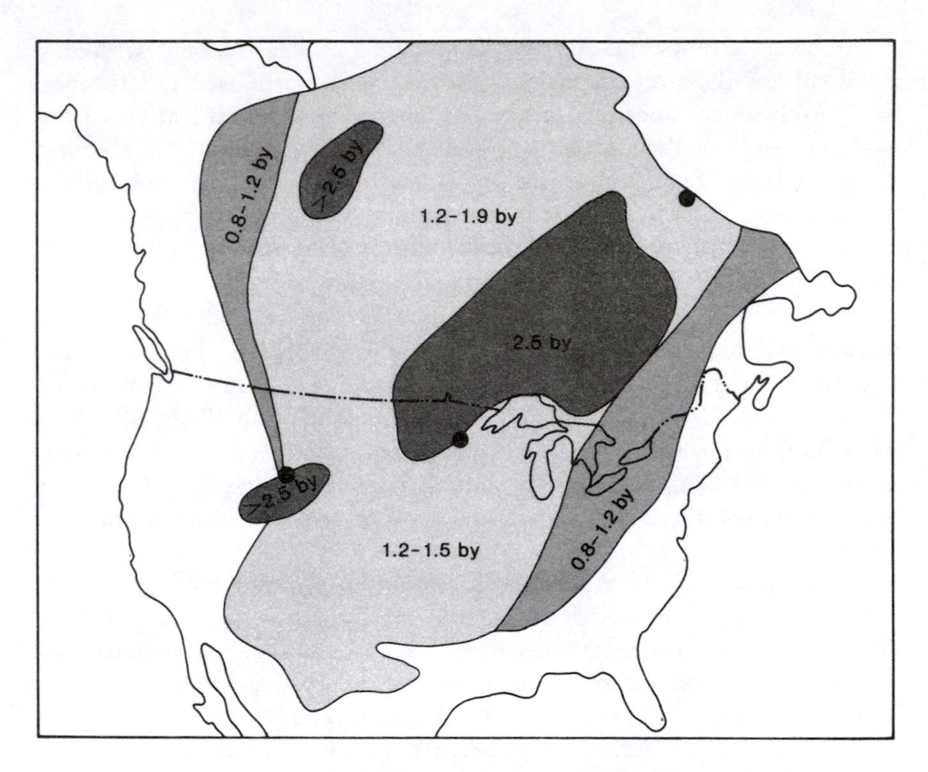

Figure 2.11. Generalized map of radiometric ages of Precambrian rocks. Black dots show places where dates older than 3.2 billion years have been obtained.

time both proceed together. These observations suggest, first, that plate-tectonically activated continental wander was going on in Precambrian time, and second, that the Grenville area was following its own course until it joined the rest of the North American continent during the Grenville Orogeny 1000 million years ago. Hairpin kinks on the main path occur at the same time interval as the Hudsonian and Grenville Orogenies, suggesting that orogenic activity may have been triggered by abrupt changes in the direction of plate motion.

Precambrian rocks underlie all parts of the United States within the dashed line shown on figure 2.13, representing the edge of the North American continent by the end of Precambrian time. Precambrian rocks east of the line from the Piedmont northward to eastern Newfoundland were probably added to the continent by plate collision during the Phanerozoic.

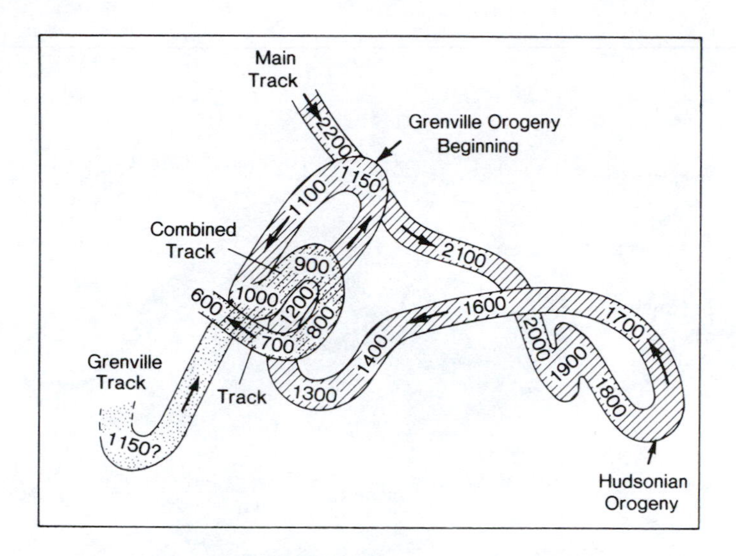

Figure 2.12. Precambrian paleomagnetic polar wander-paths for the Canadian Shield. Dates are shown in millions of years. Hairpin turns in the main track occur at the times of major orogenies. The Grenville area of eastern North America followed a separate path from the time of its earliest recognition, at 1150 million years, until 1,000 million years ago, when it joined with the rest of the Canadian Shield and the two proceeded together.

Earliest Precambrian rocks, granitoid gneisses (from which dates in excess of 3.2 billion years have been extracted using refined radiometric techniques) are known from only a few locations shown on figure 2.11. The same figure shows that Archean rocks, those older than 2.5 billion years, are not extensive south of the Canadian border. For the most part the Precambrian that underlies the central United States is granitic crystalline rock. Much of it bears the radiometric imprint of a 1350 million year orogenic event that occurred midway between the Hudsonian and Grenville Orogenies shown on figure 2.8. Of lesser areal extent are belts of Precambrian sediments and lavas mostly of Precambrian Y and Z ages (see fig. 2.8). Most of these late Precambrian bedded rocks are not much metamorphosed, retaining their primary depositional features, such as bedding and ripple-marks. Included are copper-bearing Keweenawan volcanic rocks near Lake Superior, Catoctin Greenstone and Lynchburg Formation turbidite sandstones in Virginia, Ocoee Supergroup poorly sorted shelf to deep-water clastics in North Carolina and Tennessee, and a thick sequence of sandstone, siltstone, and shale that formed in a trough extending from Death Valley, California, through Glacier Park,

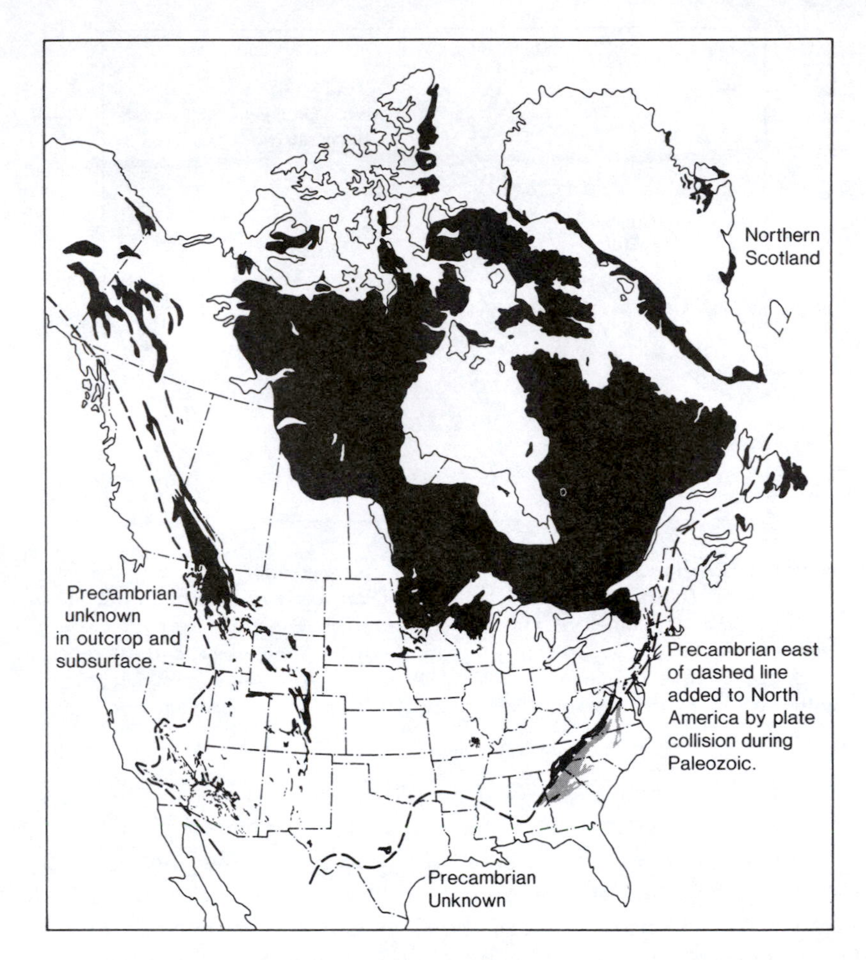

Figure 2.13. Precambrian outcrops of North America, Greenland, and northern Scotland in their restored Precambrian position. Dashed line indicates the edge of North American continent in late Precambrian time. Precambrian rocks east of the dashed line were added to North America by later plate collisions.

Montana. These are discussed in greater detail below. Along both east and west margins of Precambrian North America thick deposits of Precambrian marine rocks, principally sandstone, siltstone, and shale, are directly overlain by similar deposits of Early Cambrian age. In fact, in some sequences the Late Precambrian and Early Cambrian strata are so similar that it has been very difficult to decide where the Precambrian-Cambrian boundary actually is.

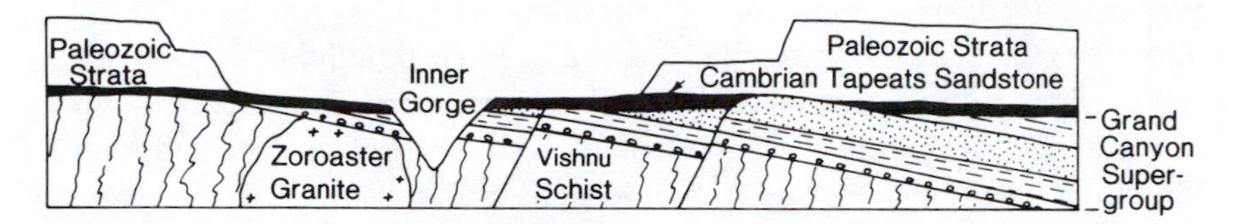

Figure 2.14. Cross section showing relationships of the Precambrian rocks at the Grand Canyon, northern Arizona.

GRAND CANYON—BELT STRATA

A remarkably continuous sequence of sandstone, siltstone, and shale was deposited in a belt 300 km wide and 3500 km long along the western margin of the Late Precambrian North American continent. The sequence is wedge-shaped. Its easternmost limit follows a sinuous line northward from central Arizona and passes just east of the Canadian Rockies; the wedge thickens westward to a maximum of 8 km at its western edge, where its record is obscured by complex Phanerozoic igneous and tectonic events.

Late Precambrian strata of this sequence are exposed at a number of well-known localities, including Death Valley, Grand Canyon, Uinta Mountains, Glacier Park, and the Canadian Rockies. The Grand Canyon section typifies Precambrian relationships in this belt. Late Precambrian strata there are assigned to the Grand Canyon Supergroup (figure 2.14). This supergroup includes the Unkar Group of Precambrian Y age and the Chuar Group of Precambrian Z age. The two groups aggregate 4 km in maximum thickness, and consist mostly of shallow-water marine siltstone, shale, and sandstone with a few thin algal limestone beds, some basalt dikes, sills, and flows, and a basal conglomerate, shown on figure 2.14. The Grand Canyon Supergroup rests unconformably on the Vishnu Schist and Zoroaster Granite of Precambrian X age. All Precambrian units have been faulted and tilted slightly. The Cambrian Tapeats Sandstone lies with angular unconformity upon a late Precambrian topography of low relief; certain Precambrian hills were too high to have been covered by the Tapeats, which is only about a hundred metres thick.

Late Precambrian sediments are grandly displayed in the Northern Rocky Mountains from Glacier Park, Montana, northward into Alberta and British Columbia. At Glacier Park they are designated as the Belt Supergroup of Precambrian Y age. Canadians call similar strata the

Purcell Supergroup. Belt-Purcell exposures constitute the greatest expanse of well-preserved late Precambrian sediments on the continent. They have been extremely well preserved in the 800 million years since they were laid down, subjected only to the lower grades of metamorphism and little folding, although they have been moved eastward scores of kilometers during Mesozoic orogenic activities.

Belt-Purcell rocks attain a maximum thickness of 15 km, not too incredible an amount considering that they represent almost a half a billion years of earth's history 1300 to 800 million years ago. Belt-Purcell sediments, like those of the Grand Canyon Supergroup, are predominantly fine-grained clastics: shale, siltstone, and sandstone. Stromatolitic algal limestone beds are common in some horizons, and basaltic lavas occur in several places, particularly along the western edge of the belt. The environment of deposition was probably near-shore shallow-water mud flats, with algal heads projecting above the water in places. Thus the late Precambrian west coast of North America more nearly resembled our modern shallow-water Gulf Coast than our present west coast!

Belt-Purcell strata are overlain with slight angular unconformity by the Windermere Group of Hadrynian (Precambrian Z) age. This group attains a maximum thickness of 5 km and includes considerably coarser sediments than the Belt-Purcell, suggesting that source areas of considerable relief developed during west coast orogenic events dated at about 800 million years before the present. Windermere sediments form a wedge that thickens westward. The original western extent of the Windermere deposits is not known, having been obliterated by Phanerozoic igneous activity along western North America.

LATE PRECAMBRIAN WORLD GLACIATION

Probably the most extensive period of glaciation in earth history occurred between 1000 and 600 million years ago. Its effects are found on all continents except Antarctica (where present glaciers prevent much searching), as shown on figure 2.15, and are more widespread than the effects of glaciations that have been recorded during Quaternary, Permo-Carboniferous, Ordovician, and early Proterozoic time.

Rocks of glacial origin are characterized by lack of sorting and bedding and are called tillites. Some marine glacial strata have laminated bedding disrupted by clasts of random sizes dumped from icebergs or glaciers. Diamictite is a general term that includes boulder beds, boulder

Figure 2.15. Dots indicate locations where late Precambrian glacial activity has been found. Dashed lines are boundaries of late Precambrian sedimentary belts. Continents are arranged in a possible late Precambrian configuration. (After Windley, The Evolving Continents, 1977, John Wiley & Sons)

clay, pebbly sandstone and mudstone, and tillite, covering the range of glacially related deposits.

Late Precambrian glacial deposits are probably not all of the same age, worldwide. Radiometric dates suggest at least three episodes of glaciation between 1000 and 600 million years ago. Within restricted areas the occurrence of a diamictite within the late Precambrian stratigraphic sequence forms an easily recognizable marker horizon that has great value in local correlation within these typically unfossiliferous sequences. For example, diamictite occurs near the base of Precambrian Z deposits throughout much of their extent in western North America, as shown on figure 2.15.

Occurrence of glacial deposits at several widely separated times throughout earth history invalidates any notion that the earth has become progressively colder since its origin. Rather it has remained remarkably constant in its temperature range and the scattered periods of glaciation represent fluctuations well within the range of modern temperature variations.

The ultimate cause of earth's glacial cycles is not presently known.

PRECAMBRIAN ECONOMIC DEPOSITS

Precambrian rocks are prolific sources of almost every kind of mineral commodity except fossil fuels. Iron, nickel, gold, silver, uranium, and chromium are all produced in important quantities. Each metal occurs within a characteristic framework, the understanding of which facilitates the search for additional ores. For example, Archean greenstone suites yield different metals from differing levels in their stratigraphic sequence: nickel, chromium, asbestos and talc typically occur in the lower ultramafic rocks; gold, silver, copper, and zinc are associated with intermediate volcanic rocks in the middle of the sequence; banded iron formations and manganese occur toward the top.

Proterozoic rocks have yielded the bulk of the world's iron, much of it as sedimentary iron oxides concentrated from the erosion of Archean greenstone terranes. Gold and uranium ores have also been concentrated in Proterozoic conglomerates from erosion of extensive granitic intrusions of earlier ages. The Precambrian embraces such a long period of time and includes so many intrusive and volcanic events that it is not surprising that this broad interval is so widely productive.

Information gathered in the search for minerals in Precambrian rocks forms the backbone of our understanding of that vast interval of time!

PRECAMBRIAN TECTONICS

Although we may never know the actual size, position, or shape of the early Precambrian (Archean) continents, we can make certain inferences about their origin and mobility from the study of radiometric dates and the structure and composition of Archean rocks. There are only a few places where rocks that retain radiometric dates older than 3.2 billion years have been recognized: Minnesota, Labrador, West Greenland, Norway, and Rhodesia. Judging from the widespread distribution of rocks of the 3.1-2.7-billion-year interval, the bulk of continental growth apparently occurred after 3.2 billion years ago.

Two or perhaps three times as much heat was generated by radioactive decay in the Archean than today. This nuclear energy, added to heat generated by formation of the core and the impact of meteorites during the Archean, promoted differentiation of the mantle and gave rise to the granites and greenstones of which the early continents were formed. It is estimated that between half and two-thirds of the present continental crust had come into being by the end of Archean time 2.5 billion years ago.

The comparatively small volumes of quartzite and carbonate rocks found in Archean rocks suggest that continental shelves were hardly developed at this time. Our picture of the early Archean is one of many microcontinental masses moving rapidly about and perhaps fringed by narrow continental shelves and island arcs. These small continental masses gradually combined to form larger continents. Conditions must have changed greatly during late Archean time, because in the early Proterozoic rock record we recognize several sequences of shallow water sediments that formed in geosynclines on continental platforms.

Continental growth and stability that was attained by early Proterozoic time resulted in the development of Proterozoic continental margins whose behavior pattern in terms of geosynclinal subsidence and igneous activity was similar to that which many continental margins have exhibited since that time. We can identify Proterozoic rock suites that are typical of various modern plate-tectonic environments and we thus conclude that plate interactions similar to those of today had been established by Proterozoic time. The plates were probably thinner, smaller, and may have moved more rapidly, but the basic plate-tectonic relationships were similar to those of today.

3

Cambrian

TOPICS

CAMBRIAN PALEOGEOGRAPHY

Visualize the face of North America half a billion years ago. Its land surface bears no living thing, plant, or animal. Nor does it resemble the moon's cratered and lifeless surface because the effects of running water are everywhere – in rills, gullies, and rivers, all sweeping weathered rock materials toward the shallow seas bordering the continent. There are no Rocky Mountains, no Appalachians. What hills there are lie along a low Transcontinental Arch extending from New Mexico to Wisconsin, where it joins the Canadian Shield. A continental framework that will be recognizable throughout several early Paleozoic periods was by earliest Cambrian time already formed. The North American interior then was a barren lowland similar in elevation to present-day Texas or Florida; offshore (in the present continental interior) was a shallow, continental shelf on which beach and near-shore sands, muds, and carbonate rocks accumulated.

Our reconstruction of Cambrian paleogeography is based entirely upon our interpretation of Cambrian rocks. Erosion has long since removed any vestige of original landscape. Nevertheless, several conclusions can be drawn about the environment of deposition of Cambrian strata in North America:

1. With a few exceptions the sediments were deposited in shallow marine waters.
2. The sediments were mainly derived from erosion of the exposed continental interior, in contrast with rocks of younger periods whose sources were often along the continental margins.
3. Thicknesses of Cambrian strata exceeding 1.5 kilometres are found only in Utah and Nevada, the Canadian Rockies, and the Appalachians, within what were in Cambrian time the unstable margins of the continent.
4. Early Cambrian deposits are limited to the continental margins, whereas late Cambrian strata transgressed across most of the United States making Upper Cambrian deposits some of the most widespread rocks in North America.

NAMING OF CAMBRIAN

The Cambrian System was named in 1835 by Reverend Adam Sedgwick as a result of his studies in Wales (Romans had called the region Cambria). At the same time, Roderick Murchison defined the Silurian System in the Wales-England border area as shown on figure 3.1. Soon

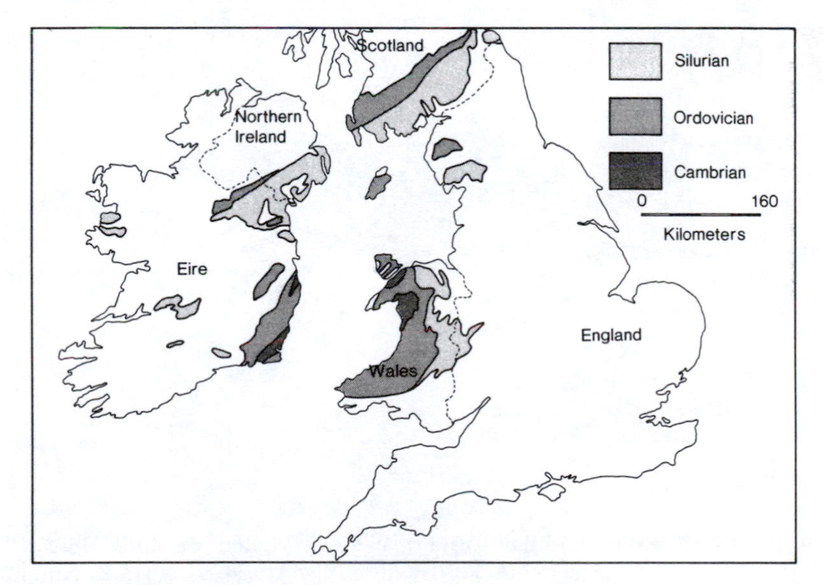

Figure 3.1. Map of southern British Isles showing distribution of Cambrian, Ordovician, and Silurian strata. The original type localities for the three systems were established as reference standards in this area.

thereafter the two men realized that their defined systems overlapped, and thus began a contest of wills to see whose name for the duplicated beds would prevail. Murchison presented a stronger case because he had taken pains to include fossils in his description of the Silurian System, whereas Sedgwick had described only the rocks. The problem was ultimately resolved by Charles Lapworth's proposal in 1879 that the intervening strata be named the Ordovician.

The Cambrian System of rocks in Wales, as defined by Sedgwick and restricted by Lapworth, has been accepted as the world standard to represent the Cambrian Period of time. In making correlations of rocks on other continents with the Cambrian of Wales, it is not feasible to compare rock types since they vary considerably; rather, all correlation is done on the basis of time which in most rocks is determined by fossils. It is unfortunate that the Cambrian of Wales is not particularly fossiliferous. This helps explain why Sedgwick did not include fossils in his original description of the Cambrian System. It also explains why sequences of more fossiliferous strata from other areas have come to be used as the references for detailed Cambrian fossil succession. Cambrian fossils have now been collected from every continent, as shown on figure 3.2.

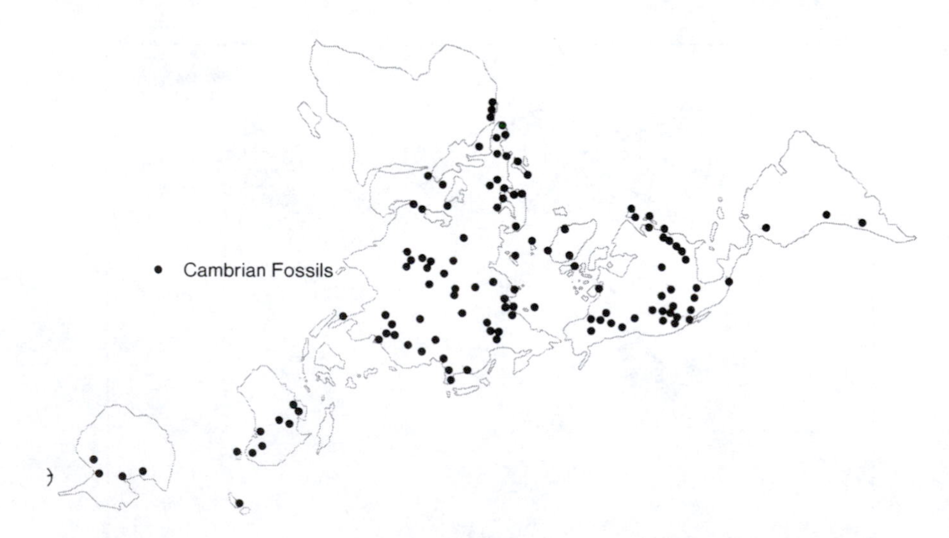

Figure 3.2. Localities in the world that have yielded Cambrian fossils. (After Palmer, Amer. Sci. 1974)

BASE OF CAMBRIAN—FOSSILS WITH HARD SHELLS

Life during the Cambrian Period was dominated by trilobites. Fossils of their segmented chitinoid exoskeletons are almost universally present in marine deposits of Cambrian age. It has been estimated that approximately 70 percent of all Cambrian fossils are trilobites. It has been common practice in the past to date the beginning of the Cambrian and the Paleozoic Era from the first appearance of trilobites like *Olenellus* (fig. 3.3). However, in many stratigraphic sequences there is a pretrilobite zone containing inarticulate (hingeless) phosphatic brachiopods that has been used by some to identify the base of the Cambrian. Russian paleontologists have recently described a prolific fauna of archaeocyathids (an extinct group believed to be transitional between colonial protozoans and sponges) from the Tommotian Stage of the Siberian Platform. Archaeocyathid appearance there is accompanied by gastropods, hyolithids, and sponges, marking the earliest preserved metazoans with hard parts.

Ediacaran-type impressions of soft-bodied organisms are widespread in Late Precambrian strata. General consensus among paleontologists appears to favor placement of the Precambrian-Cambrian boundary above strata containing Ediacaran-type fossils. Figure 3.4 shows the radiometric age of appearance of the fossil groups that are considered useful in defining the base of the Cambrian. Russian workers are making a strong case for extending the base from the earliest trilobites and

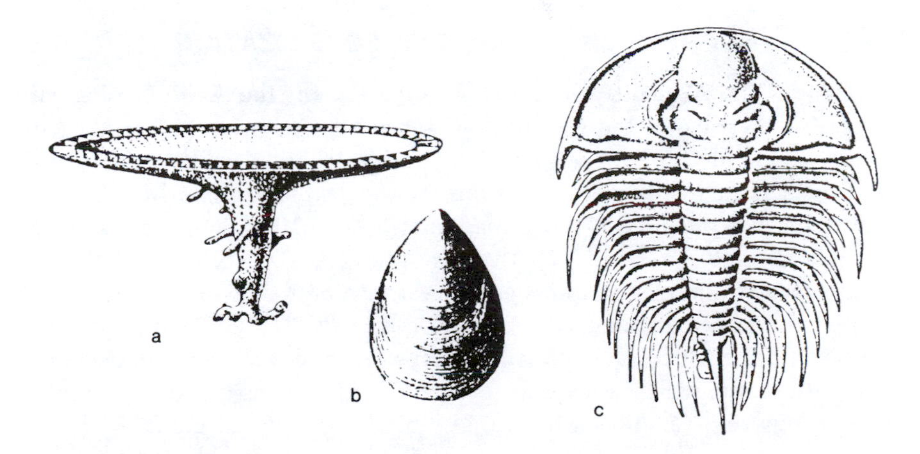

Figure 3.3. Early Cambrian guide fossils (a) Archaeocyathid, (b) Inarticulate brachiopod, (c) Trilobite **(Olenellus).**

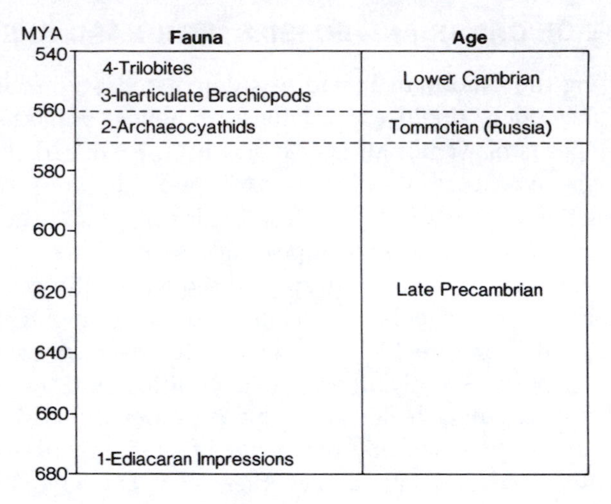

Figure 3.4. Times of appearance of major fossil groups used in defining the Cambrian-Precambrian boundary.

brachiopods down to include the Tommotian archaeocyathids. The relatively sudden appearance, about 570 million years ago, of metazoan organisms with preservable skeletons was an event of paramount importance in the history of life on earth. It may have been made possible by the development in organisms of key biochemical compounds, such as hydroxyproline, needed for the formation of collagen in body tissues.

BURGESS SHALE—UNIQUE FOSSILIZATION

Charles D. Walcott (1850-1927), director of the U.S. Geological Survey at the turn of the century and foremost Cambrian paleontologist of his day, made a most unusual fossil find with his discovery of beautifully preserved soft-bodied organisms in the Burgess Shale Member of the Stephen Formation in the Canadian Rockies, 100 kilometres west of Banff. Uniquely preserved in the fine-grained, dark-colored shale as carbonaceous films and impressions of soft structures of organisms are a number of kinds of animals – arthropods, jellyfish, echinoderms, and annelid worms. Burgess Shale trilobite fossils show soft appendages and internal organs, which are rarely preserved, giving insight into trilobite morphology and function. The Burgess Shale provides a glimpse of the teeming life of Middle Cambrian seas of which we would otherwise be unaware, and suggests that ordinary fossilization preserves only a fraction of the total fauna living at a given place and time.

CAMBRIAN FAUNAL PROVINCIALISM

Organisms are environmentally restricted; that is, certain animals are exclusively arctic, others tropical; some favor shallow, near-shore waters, others deeper, off-shore waters. So it is not surprising that Cambrian trilobites are separated into two major assemblages, which we call the "American" and "European" faunal provinces. Each assemblage is characterized by scores of trilobite genera, yet the two provinces share few genera in common.

American province trilobites are found throughout North America except along the eastern seaboard from Boston to eastern Newfoundland; and also, strangely enough, in northern Scotland. European province trilobites are found in most of Europe and along the extreme east coast of North America. Newfoundland occurrences epitomize the faunal differences. Eastern Newfoundland's European trilobite assemblages are completely different from American trilobite assemblages of western Newfoundland.

Naturally, geologists have never been at a loss for a hypothesis to explain the European-American faunal near-juxtaposition in Newfoundland. One early explanation visualized a Cambrian land mass barrier in central Newfoundland, separating two completely different basins. This idea, however, has generally been rejected and a modern explanation, shown on figure 3.5, has been proposed. Reconstructions of continental positions, as shown on that figure, put America and Europe much closer together than they are today; thus the American assemblage of northwest Scotland could have been deposited on the same continental shelf as existed in western Newfoundland.

Although the American trilobite fauna is widely distributed in North America, it is of limited distribution elsewhere in the world. The European trilobite assemblage, though absent over North America except for the easternmost edge, has a more cosmopolitan distribution throughout the world.

Trilobites in both provinces share certain evolutionary trends. One is the gradual fusion of rear thoracic segments to form an enlarged rigid tail or pygidium. Another trend is towards increasing convexity of the entire trilobite exoskeleton. Most early Cambrian trilobites are nearly flat, whereas later Cambrian forms are more convex.

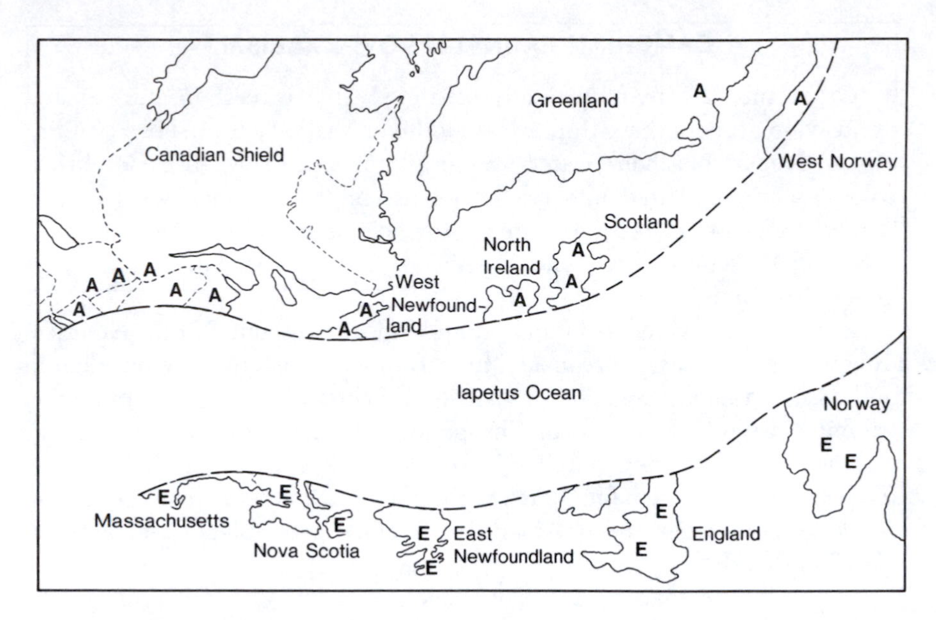

Figure 3.5. American (A) and European (E) type trilobite distributions as a means of reconstructing continental separations in Cambrian time. The Iapetus ("Proto-Atlantic") Ocean was of unknown width, but it was sufficiently wide to prevent intermingling of trilobite faunas of the same age in Cambrian and Ordovician time.

THE GREAT CAMBRIAN TRANSGRESSION

Rocks bearing Lower Cambrian fossils are areally limited in North America, although Upper Cambrian rocks are among the most widespread of any system. The spread of Cambrian seas from continental margins over the cratonic interior was not a smooth continuous event, but rather a series of transgressions and regressions that occurred throughout Upper Cambrian time.

The greatest change in inferred shoreline position occurred between the end of the Middle Cambrian (fig. 3.6A) and the earliest Late Cambrian (fig. 3.6B). If we estimate the Late Cambrian to be about 30 million years in duration, and the earliest part of Late Cambrian to occupy no more than one-third of this time, or 10 million years, then the sea's maximum advance of 1100 kilometres from central Kentucky to central Minnesota occurred at the rate of about 10 centimetres per year, hardly catastrophic! Upper Cambrian shorelines returned to their Middle Cambrian position and twice again pushed back across the continent in latest Cambrian time, as indicated on the maps of figure 3.6.

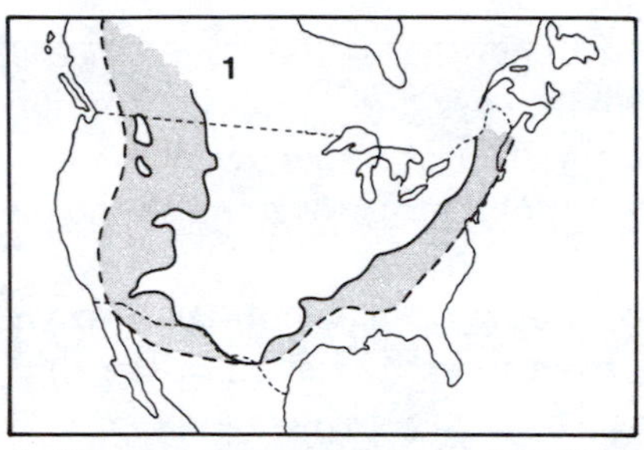

A. Shoreline at end of Middle Cambrian (1).

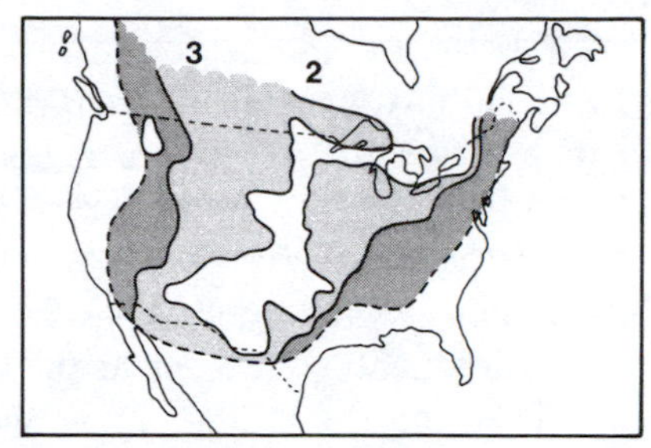

B. Early Upper Cambrian shorelines showing transgression (2) followed by regression (3).

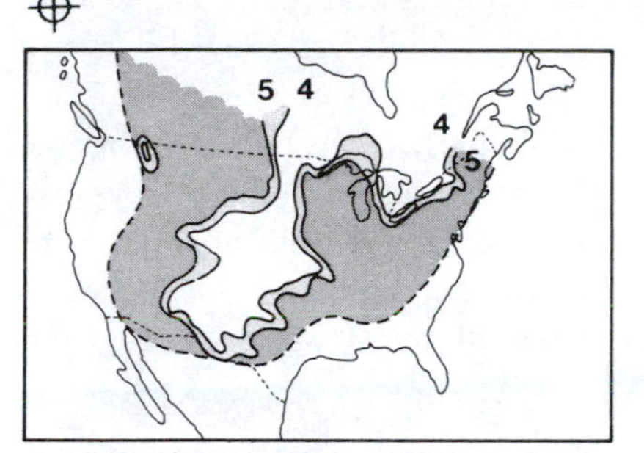

C. Middle Upper Cambrian shorelines showing transgression (4) followed by minor regression (5).

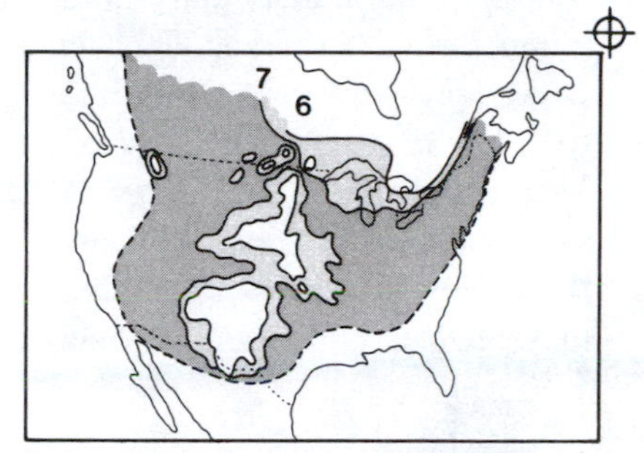

D. Late Upper Cambrian shorelines showing maximum transgression (6) followed by slight regression at end of Cambrian (7).

Figure 3.6. A series of maps of Lochman-Balk, C. (Geol. Soc. Am., Bull. v. 81, pp. 3197-3224, 1970) showing successive positions of shorelines in Late Cambrian time, as reconstructed from areal distribution of successive trilobite faunas.

CAMBRIAN GEOSYNCLINES

Three kinds of evidence suggest that the Cambrian marine waters that covered North America were no more than perhaps two hundred metres in greatest depth. The first is the paleogeography of the Upper Cambrian transgression discussed above. It seems apparent that shallow waters moving back and forth over a relatively flat continent is the best way to explain the distribution of Upper Cambrian rocks.

The second is the kind of rock of which Cambrian strata are made. Most basal Cambrian deposits are clean, well rounded, and well sorted quartz sandstones deposited on a beach or very near shore. Above the sandstones are carbonate rocks comprising the bulk of Cambrian deposits. These include bioclastic limestones, algae limestones, limestone pebble conglomerates (probably formed on tidal flats), and oölitic limestone, all of which suggest shallow water deposition.

The third evidence for shallow water deposition is the abundance of fossil life in Cambrian rocks. Trilobites with eyes and limited swimming capability would seem more at home on the floor of a shallow sunlit shelf than in the open ocean. Other associated Cambrian fossils are also indicative of shallow water environments.

Sediments and rock types from North American Cambrian geosynclines are identical, except in thickness, from those deposited in the shallow water on the craton. Thus, we conceive of a geosyncline as an area that is depressed as sediments are stacked on it in much the same way that a spring-loaded plate holder in a cafeteria maintains the same level when dishes are stacked upon it. The present delta of the Mississippi River is responding in this fashion at present. Growth of the delta is far too slow to account for the volume of sediment transported to it by the Mississippi River. The large, unseen difference is accommodated by subsidence of earth's crust beneath the delta permitting a buildup of tens of thousands of metres of shallow water sediment. The contrast between geosynclinal and cratonic thickness is apparent on figures 3.7, 3.8, and 4.3.

CAMBRIAN CONTINENTAL MARGINS

The continental interior served as the major source for Cambrian deposits, as shown by the decrease in size of sand grains and the orientation of cross-bedding away from the shield. A question remains as to the nature of the outer edge of the North American continent in Cambrian time. Evidence here is limited because later tectonism has destroyed, or later deposits have concealed, most of the critical Cambrian rocks. However, near Battle Mountain, Nevada, the archaeocyathid-bearing Scott Canyon Formation consists of 1.5 kilometres of greenstones (metavolcanics), and cherts, with little limestone. In southern Oklahoma, the Middle Cambrian basalt and andesite with associated intrusive gabbro are found between the Wichita and Arbuckle Mountains. In Colorado, small intrusives radiometrically dated as Cambrian occurred along the Transcontinental Arch. In Vermont, Cambrian rocks along the outer edge of the geosyncline have been metamorphosed by later events to schists, greenstone, and amphibolite, suggesting that they were originally, at least in part, basalt. Thus a broad outline develops in Cambrian time, of a relatively stable continental interior surrounded by less stable geosynclinal depositional belts, the outer part of which is partially made of volcanic rocks. This framework we will see repeated in other Lower Paleozoic periods.

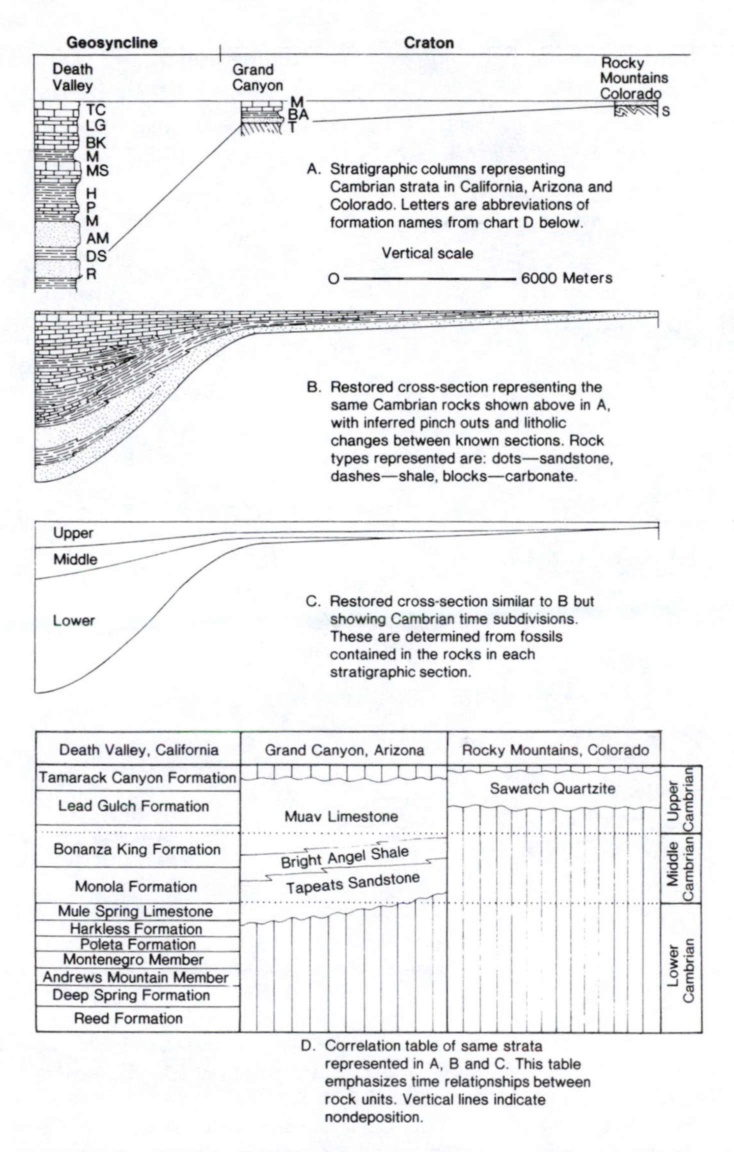

Figure 3.7. Four ways of representing stratigraphic data, each emphasizing a different aspect of the information. "A" illustrates an areal separation of the measured sections that comprise the basic data; "B" restores the probable lithologic changes which occurred between the sections; "C" shows the areal extent of the rocks at each portion of Cambrian time; "D" represents time increments by horizontal lines, emphasizing the amount of time required for each formation to be deposited but not representing the rock thickness.

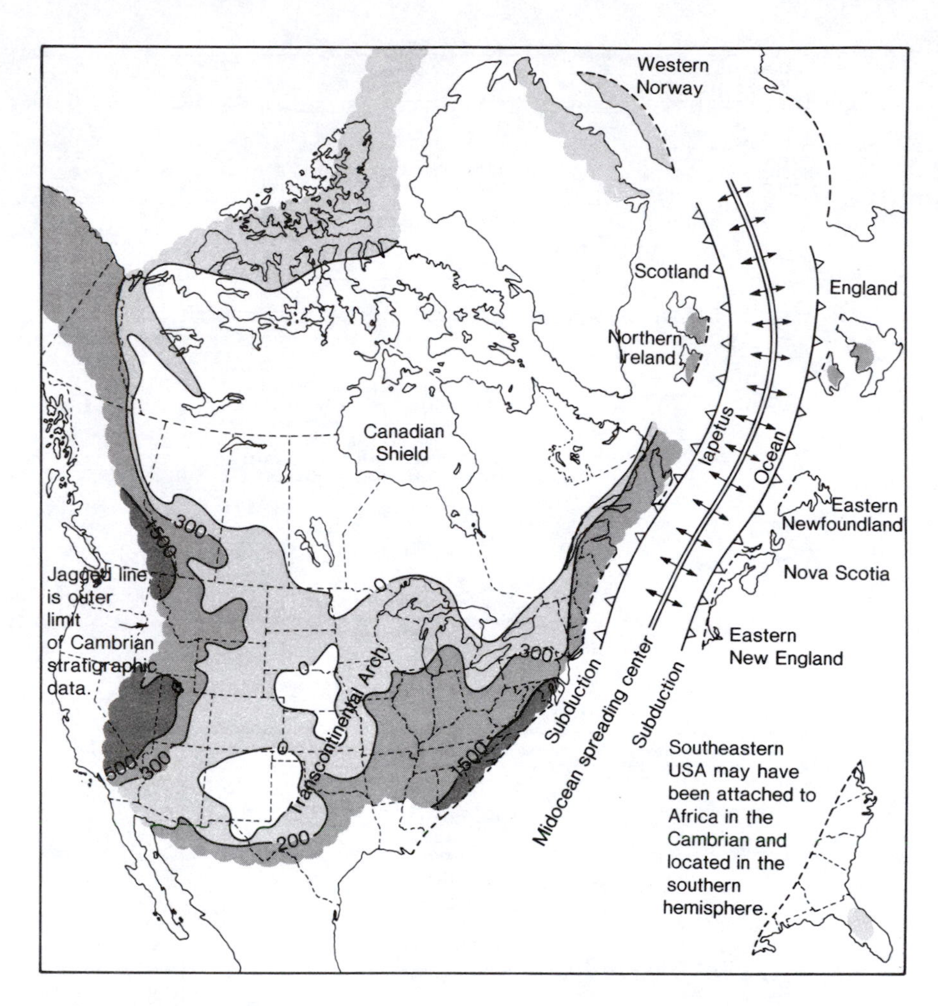

Figure 3.8. Thickness map of Cambrian System in North America. Thickness figures in metres; darkest shade show areas where Cambrian rocks are in excess of 1,500 metres (5,000 feet), intermediate shade represents thickness between 300 and 1,500 metres (1,000-5,000 feet) and lightest shade is less than 300 metres (0-1,000 feet). Transcontinental Arch and Canadian Shield served as sediment source areas throughout the period. In the continental interior other source areas existed during Early and Middle Cambrian time. Geosynclinal areas are those belts along the east and west sides where Cambrian strata accumulated to greatest thickness. The Iapetus Ocean, of unknown width, separated parts of Europe and North America. Basaltic rocks of Cambrian age in Vermont may have formed at the midocean spreading center.

CAMBRIAN TECTONICS

No Cambrian mountain-building event is indicated by Cambrian stratigraphy in North America. However, the occurrence of Cambrian volcanic rocks around the eastern, southern, and western continental margins suggests that these areas may have been the sites of Cambrian volcanic island-arc systems associated with converging plate junctions. Available evidence seems to indicate that North America was "free" during the Cambrian in the sense that it was neither connected to, nor immediately adjacent to, other continental masses. Cambrian marine transgressions indicate gradual flooding of the continent from east and west coastlines, with most of the clastic sediments being derived from the Transcontinental Arch or the Canadian Shield, low-lying areas in the continental interior. Figure 3.9 shows the paleogeography and distribution of sediments, as interpreted for the time of maximum transgression of the epicontinental seas.

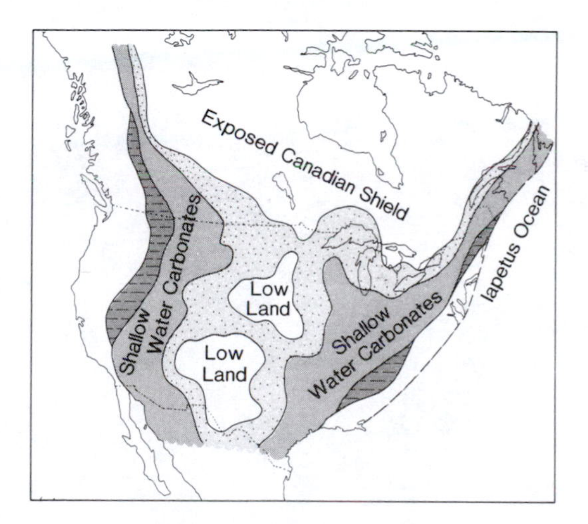

Figure 3.9. North American continent during the maximum transgression of epicontinental seas in Upper Cambrian time. The inner detrital belt (dotted) consists chiefly of beach and near-shore sands; offshore in shallow waters of eastern and western continental shelves various carbonates accumulated; an outer detrital belt of deeper water shale, limestone, and some volcanic rocks is preserved in some places on the east and west margins of the continent.

Certain Cambrian rocks still retain a faint magnetism believed to reflect the orientation of the Earth's magnetic pole in Cambrian time. Comparison of paleomagnetic pole plots of Cambrian rocks from the different continents indicates that continents in Cambrian time had different orientations and arrangements with respect to the magnetic poles than they do today. Paleomagnetic reconstructions for the position of North America during the Cambrian show the equator passing through the continent in an approximately north-south direction. This interpretation is supported by Cambrian trilobite studies that locate the warmest waters in the area postulated to be equatorial.

4

Ordovician

TOPICS

The first full-fledged North American Paleozoic mountain-making event, the Taconic Orogeny, convulsed New England and spread its deposits over eastern North America during Ordovician time. Before we discuss its complicating events, we shall outline the North American tectonic framework upon which it developed.

MIOGEOSYNCLINES—EUGEOSYNCLINES AND PLATE TECTONICS

Compare the thickness maps of Cambrian (fig. 3.7) and Ordovician (fig. 4.1) deposits. Their similarity is readily apparent. Sedimentary rocks exceeding 1.5 kilometres are found only along the outer margins of the early Paleozoic continent. However, in the Ordovician, deposits more than 1.5 kilometres thick show up in Southern Oklahoma for the first time; and the Williston Basin, now a subsurface, oil-producing basin under the northern Great Plains, shows its first thickening in the Ordovician.

We speak of the peripheral thick deposits as being laid in geosynclines. We are able to restore a more complete picture of what a geosyncline was like in Ordovician time than we were able to do for the Cambrian, even though post-Ordovician tectonic events and covering

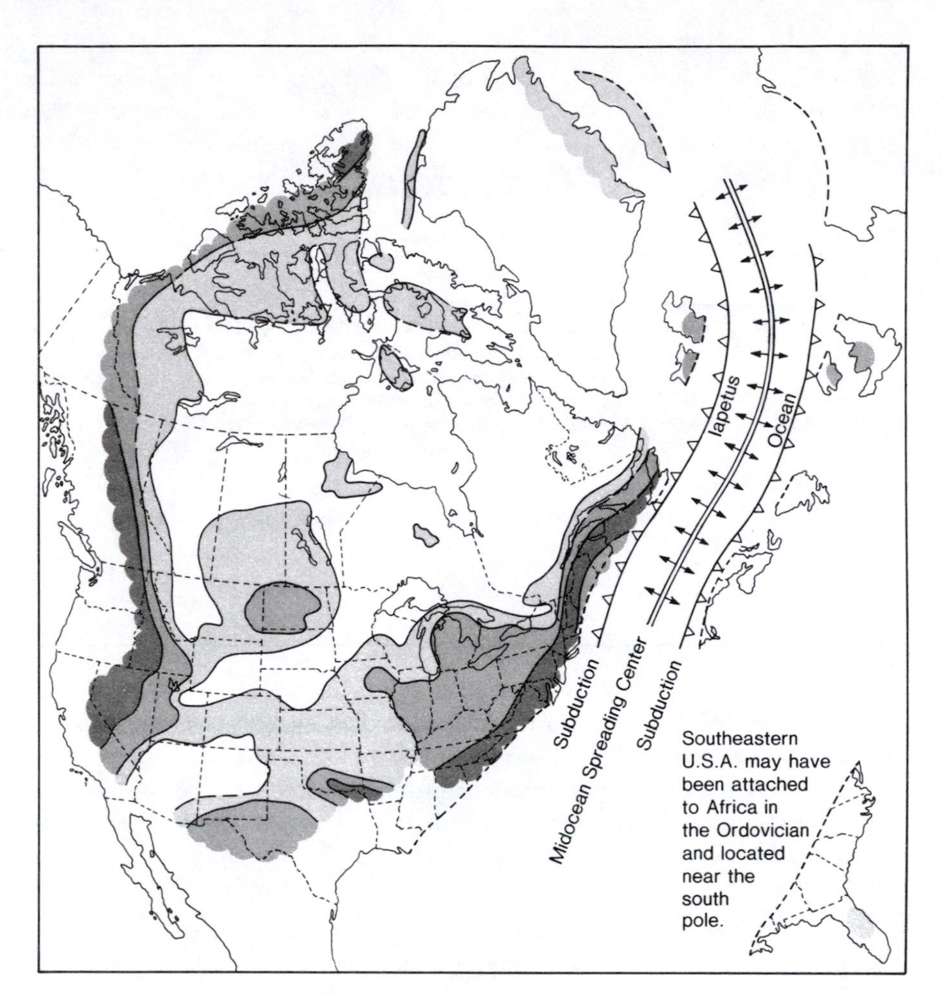

Figure 4.1. Thickness map showing extent of Ordovician System in North America as preserved today. Thickness figures in metres with thickest deposits shown by darkest shades. Greenland was closer to North America than it is today, as were the northern parts of the British Isles and western Norway. The southern British Isles and parts of northeastern North America lay on the east side of the early Paleozoic Iapetus Ocean, here shown with a midocean spreading center and subduction zones on each side. Florida at this time may have been attached to Africa and located near the South Pole. The present west coast of North America was, during the Ordovician, the site of a series of volcanic islands related to a subduction zone off the west edge of the map.

rocks still limit our view of the outer edge. If we plot the occurrences of volcanic rocks interbedded with fossiliferous Ordovician sediments, as shown on figure 4.2, we see that Ordovician volcanics were limited to the outer part of the geosyncline. Since we find this pattern repeated in rocks of many periods from the Cambrian on, it is useful to designate the volcanic rock portion of the geosynclinal belt as the eugeosyncline, and the inner, nonvolcanic portion as the miogeosyncline, as shown on figure 4.3. The volcanic rocks of the eugeosynclinal belt are commonly interbedded with shales or slates and with poorly sorted sandstones. Limestones and other shallow water sediments are rare. Fossils are less common in eugeosynclinal rocks than they are in those of the miogeosyncline and craton and both the scarcity of fossils (indicating water depths mostly greater than shallow shelf) and poor sorting of the sediments (indicating depo-

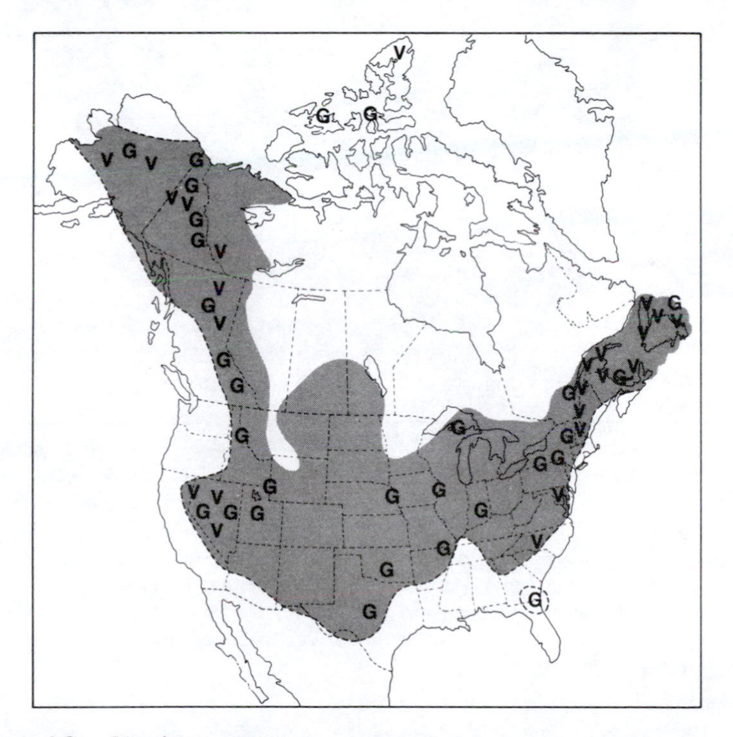

Figure 4.2. Distribution of Ordovician volcanic rocks (V) and graptolites (G). Volcanic rocks mostly represent island-arc activity or submarine flows in deeper waters around the periphery of the continent. Graptolites (free floating organisms) occur mainly in deep water shales associated with the volcanics; however, some graptolites are also found in shallow water sediments of the continental interior.

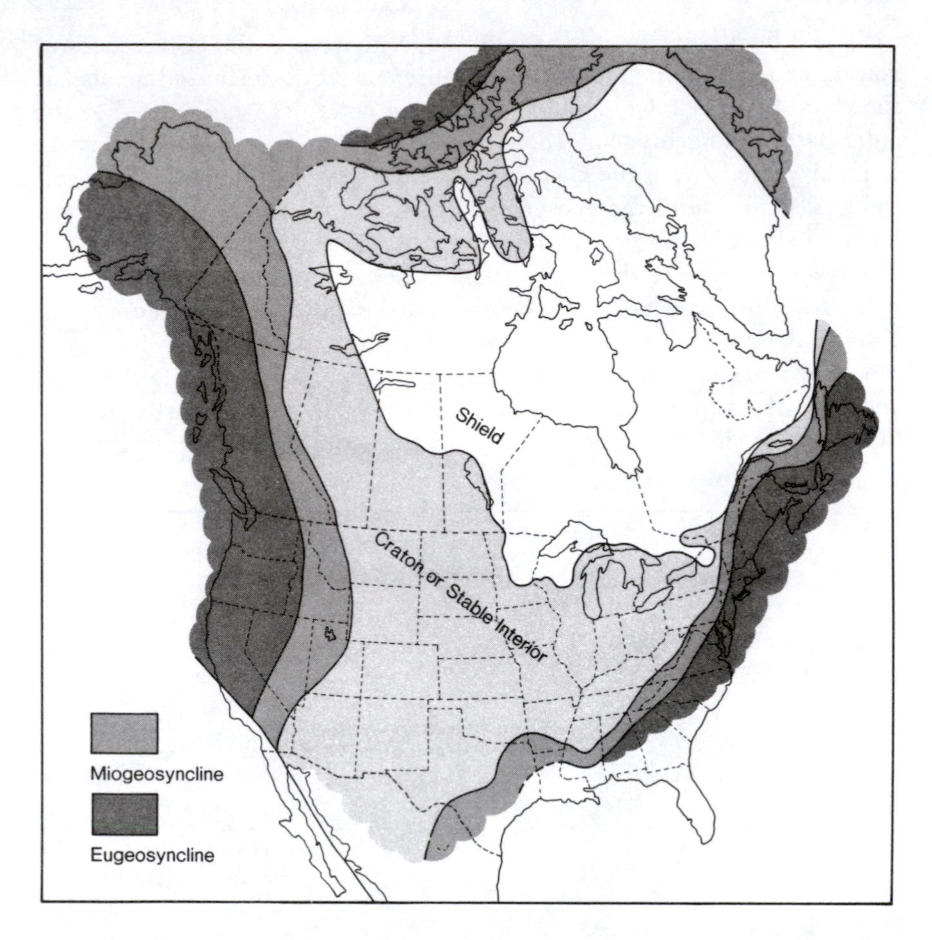

Figure 4.3. Early Paleozoic continental framework of North America. Precambrian rocks exposed in the shield were covered with a thin sheet of sediments in the stable interior or craton. Several thousand metres of predominantly shallow water, well sorted, nonvolcanic sediments accumulated in the miogeosynclinal belts. Peripheral eugeosynclinal belts were probably the sites of volcanic island arcs and subducting trench systems with associated sites of volcanic rocks and volcanic-derived sedimentary rocks deposited in moderately deep waters.

sition below wave action) suggest deposition in deeper water. A cross section of the relationships between the craton-miogeosyncline and eugeosyncline (fig. 4.4) can be compared to the present continental margin of Asia, with Japan being the offshore volcanic island arc.

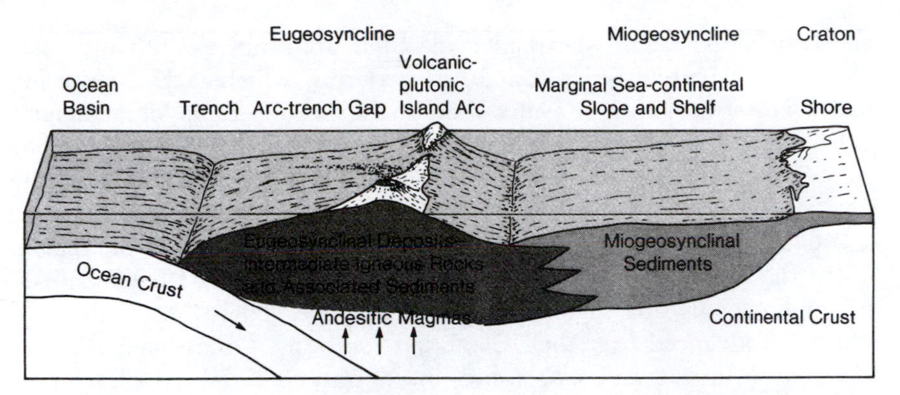

Figure 4.4. Cross section showing relations of Early Paleozoic eugeosyncline, miogeosyncline, and craton to inferred trench, subducting oceanic plate, and offshore island arc. Comparison with present continental margin of Asia, Japan Sea, and Japanese island arc is suggested.

ORDOVICIAN FAUNAL FACIES PROBLEM AND ORDOVICIAN LIFE

Graptolites are the most typical fossil group in the Ordovician. Although they appeared in the Late Cambrian and extend into the middle Paleozoic, they had their period of most rapid diversification and abundance in Ordovician time. Some graptolites were benthonic-living, encrusting objects or attached to the sea bottom, but many were floaters and lived at or just below the surface of the ocean. Because of their abundance, diversity, and wide distribution, they are almost ideal index fossils. Their chief drawback is that they are found mostly in shales, but only occasionally in limestone. Thus, graptolites are a "facies" fossil; that is, they are associated with a certain rock type, as the term "graptolite-shale" denotes. Since shales are common eugeosynclinal sediments many famous graptolite localities occur in the eugeosynclinal deposits. Here the graptolites are commonly fossilized as carbonized impressions on black shale bedding plaines, suggesting that the graptolites were preserved by settling into a quiet nonoxidizing environment. Recently it has been found that some of the best graptolite sequences can be found in thin shale interbeds within miogeosynclinal limestones, and thus make it possible to correlate the standard graptolite zones to the "shelly" fossil zones that occur in limestone.

In contrast to the limited kinds of fossils in Cambrian rocks, Ordovician limestones are filled with trilobites, brachiopods, corals, sponges, cephalopods, bivalves, gastropods, ostracods, bryozoans, echinoderms, and conodonts. Even fishes had made their appearances, although they were very primitive forms called ostracoderms, which lacked jaws and had a heavy armoring of plates around the head region. Many famous paleontologists have come from the Cincinnati, Ohio, area where they gained a boyhood enthusiasm for fossils by picking up the beautifully preserved Ordovician seashells which weather out of cliffs and roadcuts and which can also be seen on the sides of old stone houses in the region.

In the Ordovician limestone ("shelly") facies the most common useful guide fossils are trilobites and brachiopods. These forms are substantially advanced over their Cambrian forebears. Ordovician trilobites are diverse in size and shape, and, in contrast to Cambrian flat forms, most had the ability to enroll like a potato bug. Ordovician brachipods are generally much larger than the hingeless chitinophosphatic Cambrian forms, and most have hinged shells of calcareous material. Although trilobites are very abundant in Ordovician rocks of both Europe and North America, these continents have few genera in common, and intercontinental correlation on the basis of Ordovician trilobites is difficult. It is fortunate that graptolite shales are interbedded with some of the shelly faunas to make intercontinental correlation feasible. Ordovician conodonts, such as those illustrated in figure 4.5, have the attributes of good intercontinental guide fossils – common, diverse, and widely ranging. In spite of our poor understanding of the conodont-bearing animal, recently described as a simple chordate, its microscopic remains have proven to be reliable time indicators.

Reef building took a step forward in Ordovician time by including several organisms more complex than algae. Early Ordovician reefs are primarily of sponges and algae; middle and late Ordovician reefs include bryozoans and primitive corals; trilobites are common reef dwellers. Ordovician reefs are mostly small features called "patch reefs," essentially patches of clean reefoidal limestone a few metres in diameter and a metre or so in height, sitting on a muddy limestone base. Ordovician patch reefs are usually separate, not forming barrier reef structures.

ORDOVICIAN BOUNDARY PROBLEMS

It is startling to come to the realization that the commonly designated base of the North American Ordovician does not correspond with either that accepted for Sweden, or the original "type" Ordovician of Wales.

The crux of the matter is that each area has utilized different criteria for establishing the base of its own "Ordovician." Lapworth proposed the

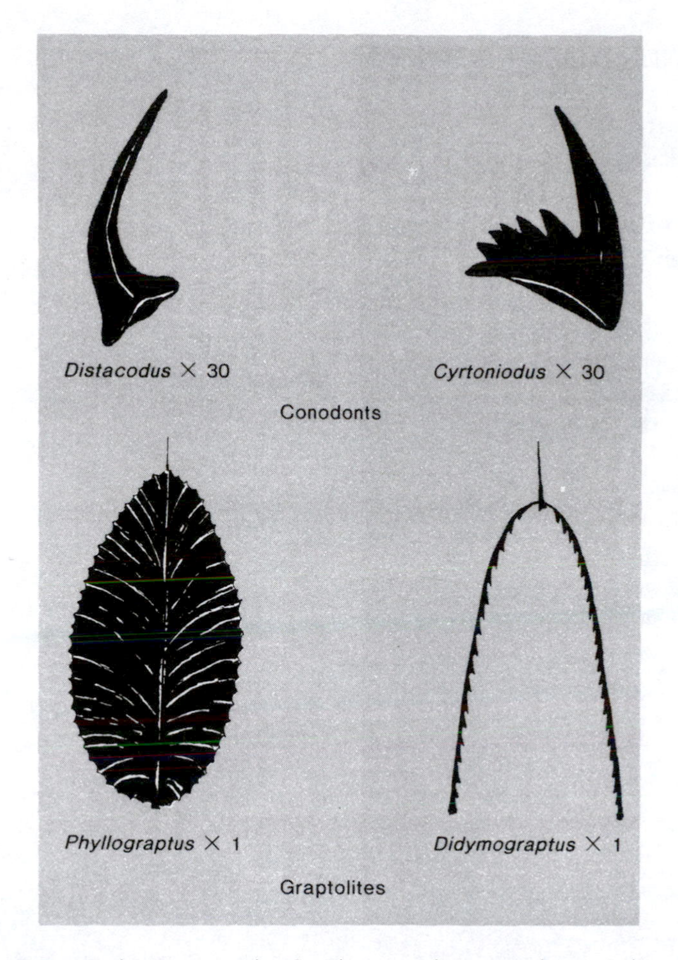

Figure 4.5. Ordovician guide fossils, conodonts, and graptolites. Both groups are extinct, but their worldwide distribution in Ordovician strata has proved to be a most useful means of establishing relative ages of these rocks.

British boundary because of an unconformity locally present at the top of the Tremadoc Series. Swedish workers noted that the Tremadoc beds contained the graptolite *Dictyonema* and were thus more similar faunally to the Ordovician than to the Cambrian; hence, they placed the boundary at the base of the Tremadoc. American workers with great areas of "shelly" facies rock specified the boundary on certain trilobite occurrences, thinking that this corresponded with the Swedish boundary. These relationships are diagrammed on figure 4.6. Recent conodont cor-

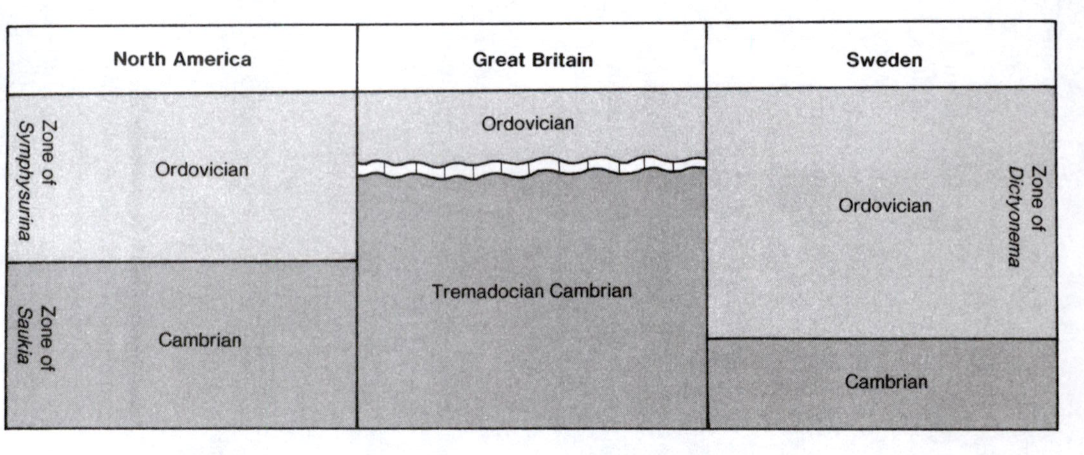

Figure 4.6. Diagram showing different usages of boundary between Cambrian and Ordovician Systems. In Sweden the Ordovician extends to the base of the occurrence of the graptolite **Dictyonema.** In Great Britain, the boundary is at an unconformity at the top of Tremadocian beds which, timewise, is several million years later than the Swedish boundary. In North America the boundary, which is between the **Saukia** and **Symphysurina** trilobite zones, has been found to be intermediate in time and between Swedish and British usages. Time on the diagram is the same along any horizontal line.

relations have indicated that the American boundary for the Cambrian-Ordovician actually falls within the Tremadoc and thus does not correspond with either the British or Swedish usage.

At one time, it was thought that geologic periods were "natural" subdivisions of time that could be recognized by finding unconformities, or breaks in the sedimentary record due to erosion, everywhere between each system of rocks. In the type areas for Paleozoic rocks, unconformities were in fact used as the basis for establishing the period boundaries. It has now become obvious that most of these particular unconformities are not worldwide in extent and that the time interval represented by an unconformity in the type area is represented by fossil-bearing rock sequences at another place in the world. Thus, fossil sequences rather than unconformities have become the practical way by which geologic units can be correlated between continents. Fossil occurrences are well enough defined now that groups of paleontologists are presently working to select zones that can most easily and accurately be used for time boundaries on an international basis. Rather than being "natural" breaks in time marked by profound worldwide unconformities, most period boundaries are arbitrary points in a continuum and are defined by fossil changes.

GLOBAL UNCONFORMITIES—TECTONIC AND GLACIAL

Having pointed out above that the early geologic system definers were not particularly successful in picking worldwide unconformities for our period boundaries, this does not mean that there were not times when worldwide sea-level maxima and minima existed. Today we have a great deal more information at hand on stratigraphic sequences and have studied them not only at the surface but have traced them into the subsurface by seismic means as well as by drilling for oil and gas. We have dated them not only by fossils but also radiometrically. Sophisticated computer analyses of worldwide data by oil company geologists have led to recognition of eustatic (meaning worldwide) changes in sea level as shown on figure 4.7.

What can cause a worldwide change in sea level? The most obvious explanation that comes to mind is to remove part of the water from the ocean by stacking it up on the land as glacial ice. This occurred most spectacularly in Pleistocene time when sea levels were as much as a hundred metres lower than today when Pleistocene icecaps were at their maxima. Conversely, worldwide sea levels would rise nearly a hundred metres if all the ice in present-day Antarctica and Greenland were to melt.

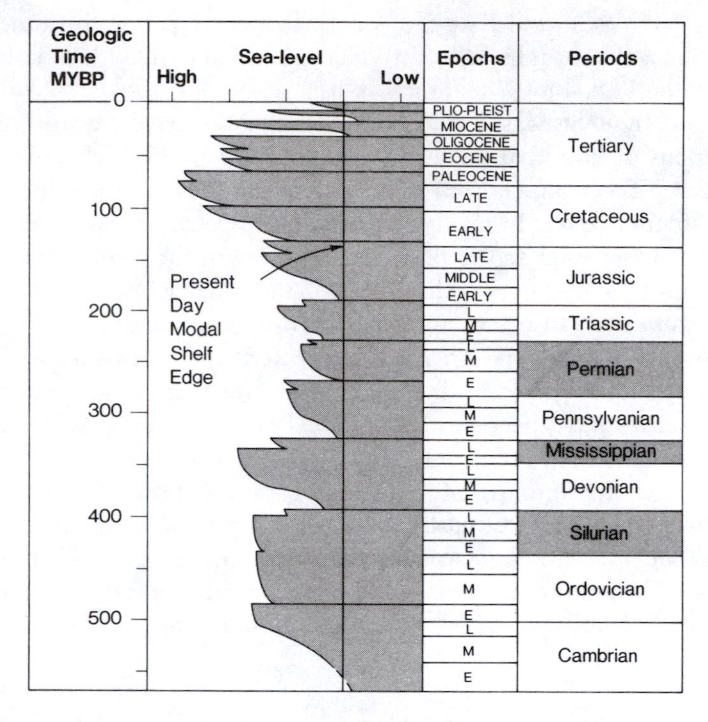

Figure 4.7. Major global unconformities occur just after sea level maxima and are indicated by the horizontal lines in the shaded sea level column. Comparison of these lines with the geologic period boundaries shows that they correspond only at the top of the Silurian, Mississippian, and Permian. (After Vail and Mitchum)

Did such a thing happen during the Ordovician? Probably so, because abundant evidence shows that continental glaciers covered parts of Africa, South America, and Europe in Late Ordovician time. The resulting drop in sea level has been used to explain differences between latest Ordovician and Silurian faunas, as well as to explain subaerial weathering features found in Late Ordovician redbeds in the eastern United States. This glacially-caused modification in sea level shows on figure 4.7 as the small nick in the sea-level graph at the Ordovician-Silurian boundary.

Another cause of sea-level change may be related to plate tectonics. Although it is not known why changes in the rate of sea floor spreading might occur, it is apparent that if there is an increase in the spreading rate there will also be a correspondingly greater uplift along the midocean ridge system, displacing the sea water, which would thus become spread more broadly on continental platforms during times of rapid sea floor spreading. Continental collisions or changes in direction of movement of major plates may be the trigger for changes in spreading rates. We do not have enough information on such events during the early Paleozoic to ascribe sea-level changes to a particular event. However, it is likely that the sea-level maximum shown on figure 4.7 at the end of the Early Ordovician was related to a plate tectonic cause rather than to glaciation.

THE TACONIC OROGENY AND PLATE TECTONICS

An orogeny is a series of mountain-making events and the Taconic is the earliest Paleozoic orogeny to affect North America. Since the Taconic involved many fossiliferous rocks, its sequence is easier to reconstruct in some detail than any Precambrian orogeny. The mountains produced during the Taconic Orogeny have long since worn away. How then do we know they once existed? What can we deduce about their location and growth?

Two major kinds of evidence substantiate the Taconic Orogeny:

1. Structural evidence of disturbance within the uplifted and deformed landmass itself.
2. Large volume of clastic sediments derived from the uplifted area and shed onto adjacent areas.

More than 100 years ago geologists recognized that a marked angular unconformity of nearly 90 degrees existed in places between folded Middle Ordovician strata and overlying Silurian beds along the Hudson River valley. Since that time unconformable relationships have been discerned at many places from eastern Pennsylvania to Maritime Canada, and the missing beds (hiatuses) at these unconformities range from a small part of the Upper Ordovician to as much as Cambrian through Silurian.

The Taconic Orogeny was not a single simple event, but rather a series of complex disturbances that began in the Middle Ordovician and extended into Early Silurian time, affecting eastern North America from the central Appalachians to Newfoundland, but having their most conspicuous effects in New England.

The Taconic Orogeny developed upon what had been the Cambro-Ordovician miogeosynclinal-eugeosynclinal belt. Most of the tectonic activity occurred within the volcanic eugeosynclinal belt, but other effects, reflected both in unconformities and in surges of classic sedimentation, spread westward across the miogeosynclinal belt and onto the craton. Structural evidences within and peripheral to the disturbed belt are varied and numerous, attesting to the mobility of the crust in this area during the Middle and Late Ordovician. These evidences follow:

1. Widespread unconformities resulting from Middle and Late Ordovician uplift and folding.
   Lower Ordovician deposits are found in both the miogeosynclinal and eugeosynclinal belts indicating normal sedimentation during Early Ordovician time. However, in the eastern part of the miogeosyncline and in most of the eugeosyncline there exists a widespread break in the continuity of sedimentary deposition during Middle Ordovician to Early Silurian time, as shown on figure 4.8. In this area Middle and Upper Ordovician and Lower Silurian beds are generally missing, and Late Silurian or even Early Devonian rocks

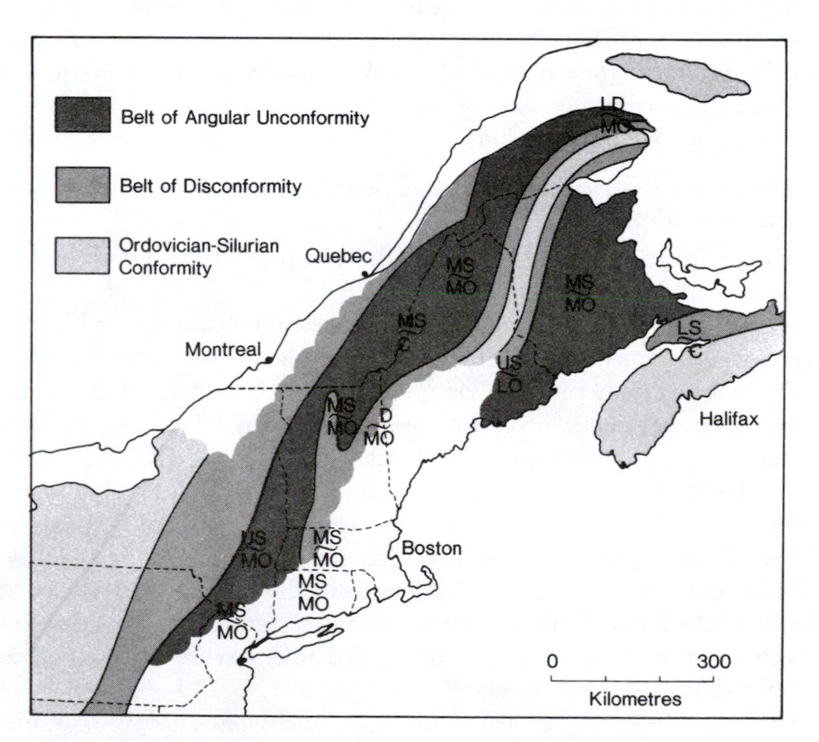

Figure 4.8. Distribution of unconformities ascribed to the Taconic Orogeny and their relationship to areas of uninterrupted sedimentation (conformity) during this orogeny. (After Pavlides, Boucot and Skidmore)

rest directly on Lower Ordovician and earlier strata in an unconformable relationship that serves as stratigraphic evidence of the uplift of the Taconic paleomountains.

2. High-angle faulting of early Middle Ordovician age in the eastern part of miogeosyncline.
   High angle, north-south faults in the eastern edge of the miogeosyncline in New York and Vermont displaced Middle Ordovician rocks over one kilometer, bringing them in places against Precambrian strata. Although it is impossible to make the connection directly, it is possible that paleoearthquakes associated with this faulting may have triggered submarine landslides that caused Ordovician and Cambrian limestone blocks of the carbonate shelves to slide into the graptolite shales accumulating in adjacent deeper water.
3. Numerous occurrences of large transported limestone blocks in Ordovician shales, probably resulting from submarine landslides.
   These limestone boulder deposits with some of the "boulders" several metres in length, much too large for normal stream deposits, occur in scattered places in eastern New York, Quebec, and Newfoundland as unique testament to instability along the edge of the carbonate banks.
4. Thrust fault emplacement of Ordovician rocks now exposed in Taconic Mountains of New York, Vermont, and Massachusetts.
   Taconic eugeosynclinal rocks now sit as a great tectonically transported mass on top of the miogeosynclinal rocks of nearly the same age in the present day Taconic Mountains in the New York, Vermont, and central Massachusetts region. Taconic strata now covering over 6,500 square kilometres were originally deposited 80-120 kilometres east of their present position, and moved westward as a series of six thrust plates during a relatively short time during the Middle Ordovician. These relationships are shown on figure 4.9.
5. Increase in volcanic activity compared to Cambrian.
   Volcanic rocks are a much more conspicuous element of the eastern geosyncline in the Ordovician than they were in the Cambrian. In additional to eugeosynclinal lavas, there are a number of widespread ash layers in the Middle Ordovician of the miogeosynclinal belt, probably derived from explosive volcanoes to the east. The ash layers, being instantaneous blanket deposits, form useful correlation horizons within miogeosynclinal limestone sequences. The surge of volcanism also indicates the increased tectonic activity in the area of the Taconic Orogeny.

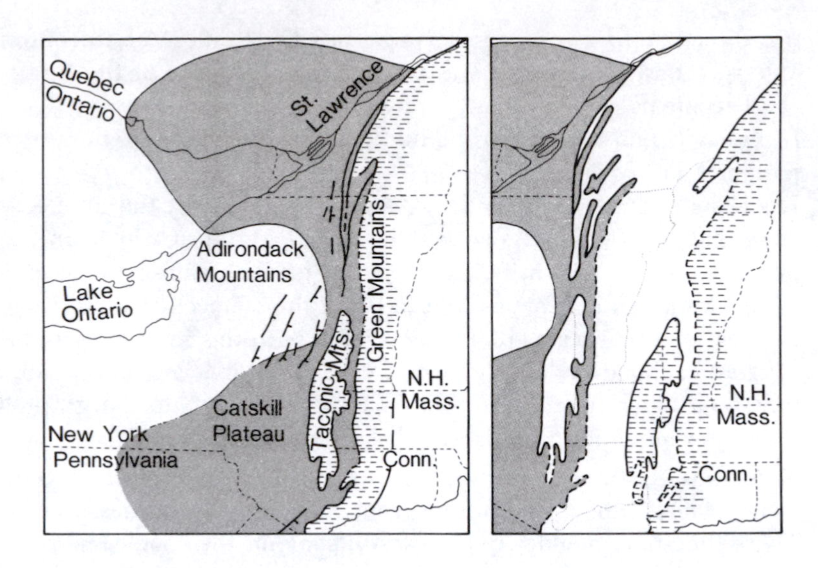

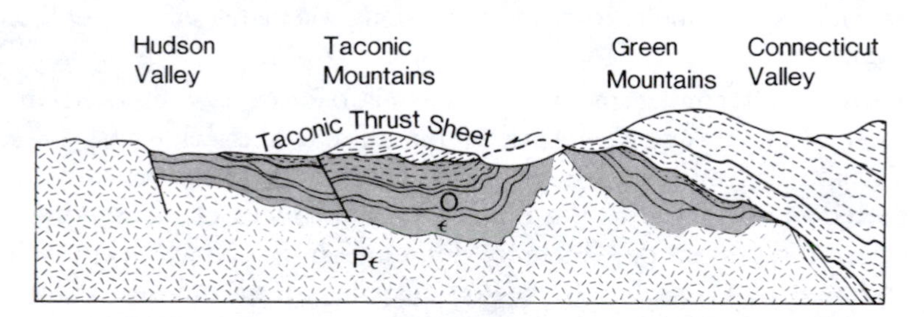

Figure 4.9. Distribution of Lower Ordovician facies in New England. The map on the left shows the present distribution as a result of Taconic thrusting; the map on the right shows the inferred distribution before thrusting. Shading represents Cambrian and Ordovician carbonate sediments; horizontal dashed pattern represents contemporaneous shales. The line of juxtaposition of differing yet contemporaneous facies was recognized more than a century ago by Sir William Logan and is often referred to by geologists as "Logan's line." Cross section shows the eugeosynclinal shales overthrust on the miogeosynclinal carbonate section. (After Kay, 1942)

6. Granitic intrusions in New England and Maritime Canada and thermal peaks in the Blue Ridge-Piedmont area to the south with isotopic dates centering around 440 million years.

   Granitic intrusions that accompanied the Taconic Orogeny have been difficult to identify – although several have been recognized – since they occur in the same general region as the more widespread plutons of Devonian and younger ages. In the Blue Ridge and Piedmont areas of the southern Appalachians, radiometric age determinations indicating periods of crystallization have shown crustal heat peaks at 1,150 million years, 800-900 (Precambrian), 440 (Ordovician), 340-360 (Devonian-Mississippian), 260 (Permian), and 180-190 (Triassic). Repeated periods of regional heating have occurred in this area and one of them coincides with the Taconic Orogeny.

7. Ultrabasic intrusives scattered from Newfoundland to Georgia, many dated at about 480 million years.
   Ultrabasic rocks were emplaced in the eugeosyncline during the early stages of the Taconic Orogeny when the geosynclinal sediments and volcanic rocks were being folded. The ultrabasic rocks now exposed at the surface were probably intruded into eugeosynclinal materials and solidified at various depths as they rose through the rocks. Some ultrabasic rocks are interpreted to indicate ancient positions of volcanic island arcs, while others are regarded as relicts of midocean spreading centers of the Iapetus Ocean floor.
8. Metamorphism of most early Paleozoic rocks in the eastern eugeosynclinal belt.
   Metamorphism pervades the rocks of New England, the Ordovician eugeosyncline, to such an extent that for many years these rocks were all thought to be Precambrian shield gneisses and schists. Finding fossils within certain metamorphic rocks led to the realization that these metamorphosed strata were mostly Paleozoic in age. The most severe metamorphism in this area occurred in the Devonian, and it is difficult to read through these later events the degree and extent of metamorphism that accompanied the Ordovician Taconic Orogeny. Nonetheless, radiometric ages of 440 million years in certain schists in Vermont and study of rotated garnet crystals in which two periods of deformation are preserved, have led some observers to conclude that much of the regional metamorphism in the New England area may be attributed to the Taconic Orogeny.
   As the eugeosyncline was folded into the Ordovician Taconic paleomountains along the eastern coast of what is now Maryland to Maine, they shed sediments westward over an increasingly larger area. First evident in late Middle Ordovician as the Schenectady Sandstone and Martinsburg and Canajoharie Shale Formations in the miogeosyncline of eastern New York and Pennsylvania, sediments were transported as far westward as Michigan and Ontario. Known as the Queenston red

shale, or sometimes called the Queenston delta complex, as shown on figure 4.10, the sediments are mostly shale and are thickest and coarsest along the eastern edge of the miogeosyncline, indicating westward transport. Their volume is impressive and reflects the uplift and erosion of a substantial part of the continental margin and adjacent ocean basin. The area remained a major source into the Silurian time when it was finally worn down.

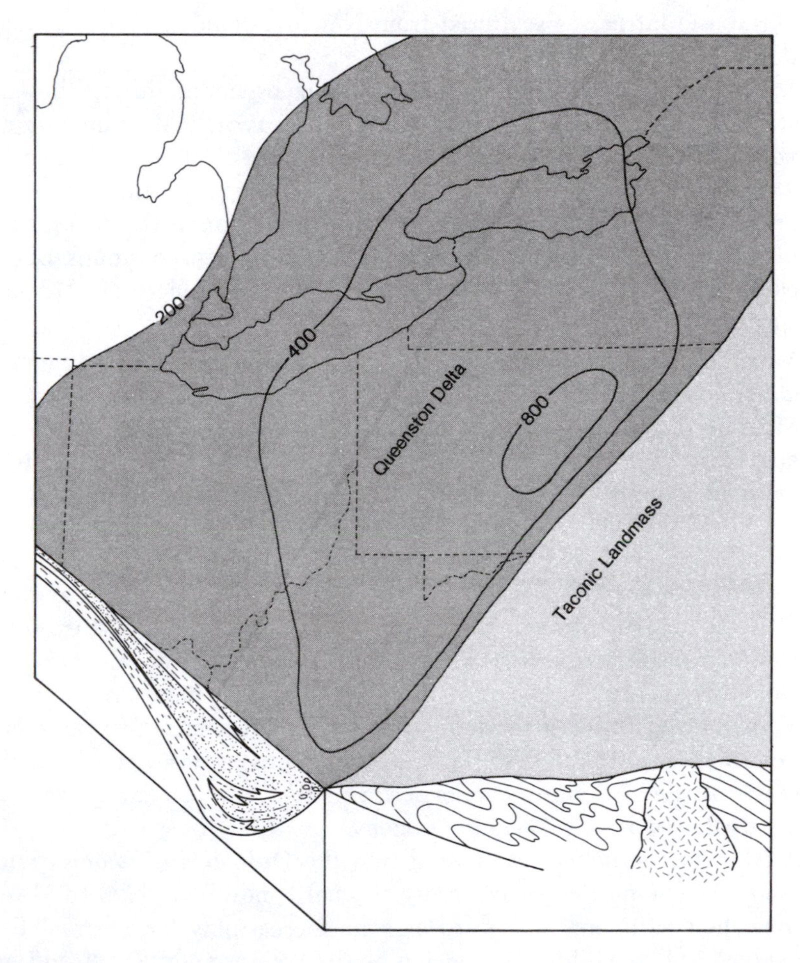

Figure 4.10. Thickness map (figures in metres) and cross section of Upper Ordovician Queenston delta complex. Sediments are coarse sandstones and conglomerates immediately adjacent to the Taconic orogenic landmass; these pass westward into the red shales and mudstones that characterize most of this ancient deltaic complex.

EQUATORIAL POSITION OF NORTH AMERICA DURING ORDOVICIAN TIME

We do not have an unequivocal fix for the global position of North America during Ordovician time. Reliable paleomagnetic information on Ordovician rocks is scarce, but suggests that North America lay at low latitudes and was transected by the Ordovician equatorial belt. Three estimates of the trend of the Ordovician equator are shown on figure 4.11. Lest the variation of these trends seem too alarming, we must realize that paleomagnetic data give only relative latitude from the pole and that the longitudinal trends must be estimated from other data, such as the paleogeographic distribution of fossils and salt deposits. Certain fossil

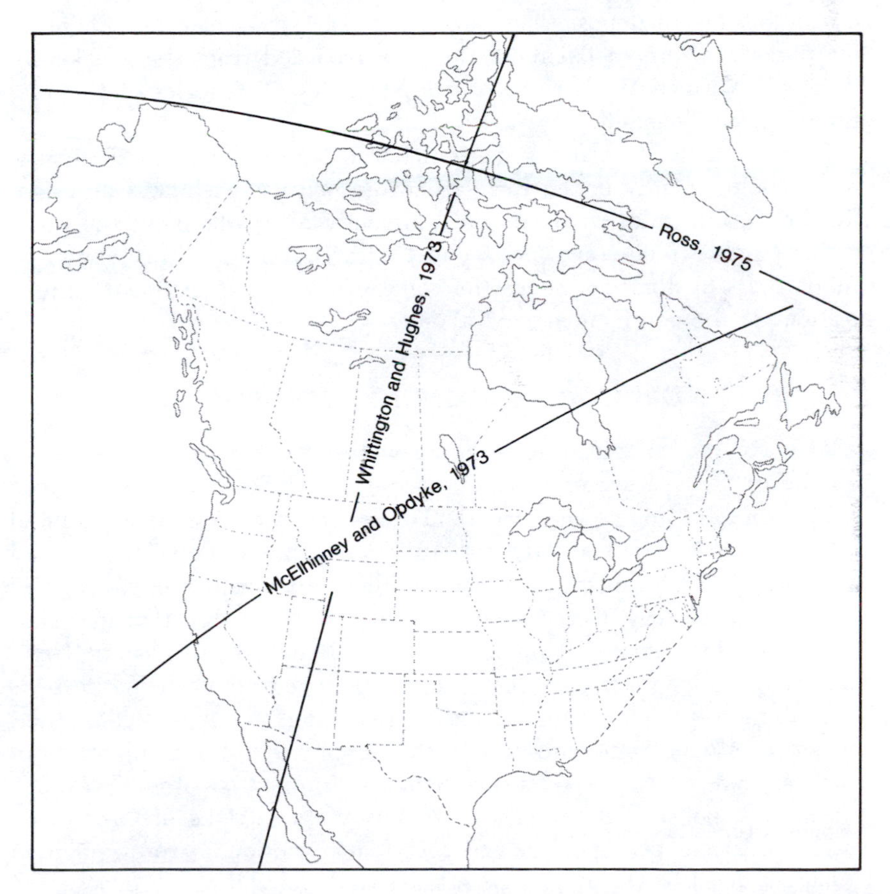

Figure 4.11. Position of North America relative to the equator during Ordovician time as given by three different groups of authors.

faunas, by analogy with their modern counterparts, are judged by paleontologists to represent warm-water organisms, while other faunas are deemed to have lived in cooler oceans. The occurrence of warm-water faunas of Lower Ordovician age in Utah, British Columbia, and Greenland supports paleomagnetic data placing North America in an equatorial position. Florida, on the other hand, has Ordovician faunas similar to those in North Africa that are thought to have lived in cool waters. This is substantiated by evidence for Ordovician glaciation of Africa; paleomagnetic data suggest that Africa (and Florida) were at the South Pole during the Ordovician, a real switch for the Sunshine State.

Evaporites are useful as indicators of climate, generally being taken to indicate the same latitudes north and south of the equator as deserts occupy today. Salt deposits have been found in strata of every Phanerozoic period; in Ordovician rocks they are recorded from the Mackensie Basin and Norman Wells areas of the Northwest Territories of Canada, and also from Thailand.

Paleomagnetic data also suggest that the North American Plate was moving quite rapidly in the early Paleozoic because the few Cambrian, Ordovician, and Silurian polar plots currently available show the poles of these ages widely separated. The Ordovician Period was about 60 million years in duration, hence the equator must have occupied various positions as the continent moved about.

ORDOVICIAN ECONOMIC DEPOSITS

Most North American mineral resources of Ordovician age are a product of shallow water marine sedimentation. Petroleum deposits in Ordovician sandstone, limestone, and dolomite beds are of substantial importance in Oklahoma and northern Texas. The St. Peter and related sandstones, widely distributed by Middle Ordovician regression and transgression of seas over the midcontinent, are valuable raw materials in glass-making. The Wabana iron ores of Belle Isle in eastern Newfoundland, no longer in production, came from red oolitic hematite beds up to 29 metres thick. Ordovician marble is among the most widely used for facing stone. Some are true metamorphic marbles from areas in Vermont that were heated during Paleozoic orogenies; other "marbles" are, in fact, polished limestones that show cross sections of Ordovician brachiopods and corals as well as secondary solution features, such as stylolites. Because Ordovician limestones and dolomites are exposed at the surface over a large area in the eastern United States, they are widely used for crushed stone for roads and concrete aggregate, burnt for lime and cement, and used for flux in the reduction of iron ores.

5

Silurian

TOPICS

Silurian was the shortest Paleozoic period. Radiometric dates indicate it embraced less than half of the time included in either the Ordovician or Devonian. Nevertheless, the Silurian System in North America is unusual not only in containing commercial iron ore, salt, and gypsum deposits, but also in its tectonic behavior that made possible extensive salt deposits in the north-central United States. Silurian deposits now cover a substantially smaller part of North America than do either Cambrian or Ordovician strata. This can be seen by comparing their distributions as shown on figures 3.8, 4.1, and 5.1. However, the same general pattern of distribution observed in the earlier Paleozoic persists, with thickness exceeding 1.5 kilometres only in the marginal geosynclines.

FINAL DEPOSITS FROM TACONIC MOUNTAINS

The waning depositional effects of the Taconic uplift are clearly visible on the Silurian map and section (figs. 5.1 and 5.2). A lens-shaped deposit, with its coarsest sediments adjacent to the source, is 1.5 kilometres thick in eastern Pennsylvania. The basal Silurian quartz sandstone, called the Tuscarora Sandstone, is widespread in the folded Appalachians, where it is a prominent ridge-former. Paleocurrent structures and grain size distributions show that the Tuscarora Sandstone was derived

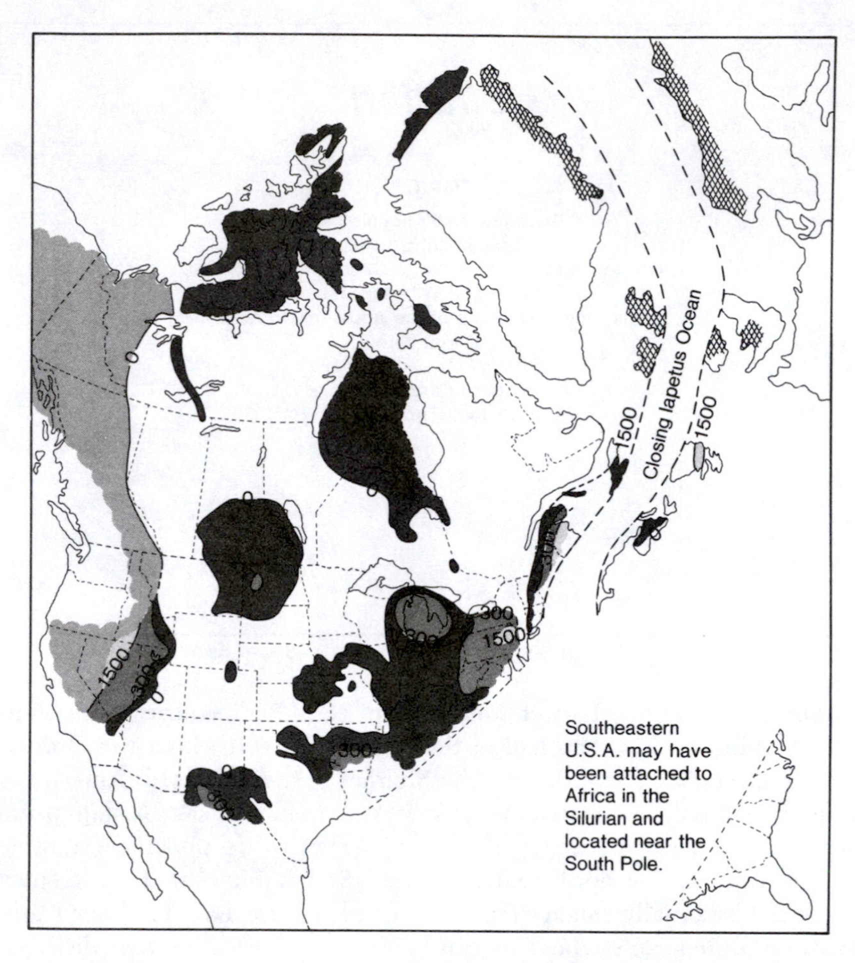

Figure 5.1. Thickness map of Silurian System in North America. Thickness expressed in metres. The Iapetus Ocean was closing during Silurian time, bringing North America closer to Europe and resulting in less faunal provicialism than shown by Cambrian and Ordovician faunas. The cross hatched areas were affected by the Caledonian Orogeny.

from a source terrane located east of present outcrops. The Tuscarora and related coarse-grained basal Silurian clastic rocks total more than 100,000 cubic kilometres in volume attesting to the magnitude of the Taconic source. Succeeding redbeds form a deposit sometimes called the Bloomsburg "delta," a Silurian counterpart of the Ordovician Queenston "delta," both products of the Taconic Orogeny. These are not deltas in the strict sense. They are wedges of clastic deposits, principally shales, siltstones, and sandstones, that accumulated under a variety of depositional conditions,

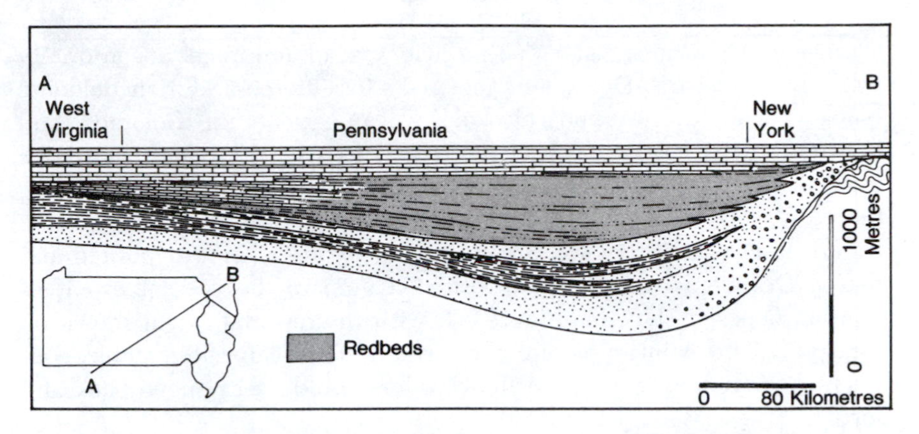

Figure 5.2. Cross section through Bloomsburg "delta," sandstones and shales derived from Taconic orogenic belt. Late Silurian carbonates at top of section indicate that the Taconic highland had worn down or subsided by the end of the Silurian.

some of them deltaic, in the subsiding basin along the western edge of the Taconic sourceland. This kind of deposit has been termed an exogeosyncline. By the close of the period, carbonate sediments were deposited in this area indicating that the Taconic Mountains were no longer an effective sediment source in northeastern United States. Maritime Canada also includes areas that received substantial Silurian deposits. Hundreds of metres of clastic sedimentary and volcanic rocks accumulated throughout Silurian time in narrow troughs adjacent to volcanic uplifts within the eugeosynclinal belt in eastern Canada. Neither granitic intrusions nor widespread folding episodes have been identified in Silurian time in North America; however, a major orogenic event, called the Caledonian Orogeny, is recorded in Silurian and earlier strata in parts of the British Isles, Scandinavia, and East Greenland as shown on figure 5.1. Some writers describe the Caledonian fold belt as a northward continuation of the earlier Taconic deformation.

CONTINENTAL STABILITY

Most North American Silurian deposits, excluding those derived from the Taconic uplift and those in the volcanic eugeosynclinal belt, are part of a great carbonate blanket of dolomite and limestone. We think this carbonate sheet was probably much more extensive as originally deposited than as now preserved, for the following reasons: (1) the present edges of the Silurian carbonates do not exhibit facies changes characteristic of a depositional edge; (2) a widespread erosional unconformity exists beneath early Devonian strata; (3) fossiliferous Silurian dolomite blocks collapsed into mid-Paleozoic volcanic vents in Colorado and Wyoming, and have thus been preserved in an area where Devonian erosion has removed all other traces of Silurian strata. We conclude, therefore, that Silurian carbonates once covered most of the North American continent, but were removed shortly after being deposited and before Devonian strata were deposited. Compare the present extent of Silurian deposits, shown on figure 5.1, with the restored extent shown on figure 5.3. This paints a picture of a virtually flat, featureless, stable, continental interior over which shallow waters could easily move back and forth.

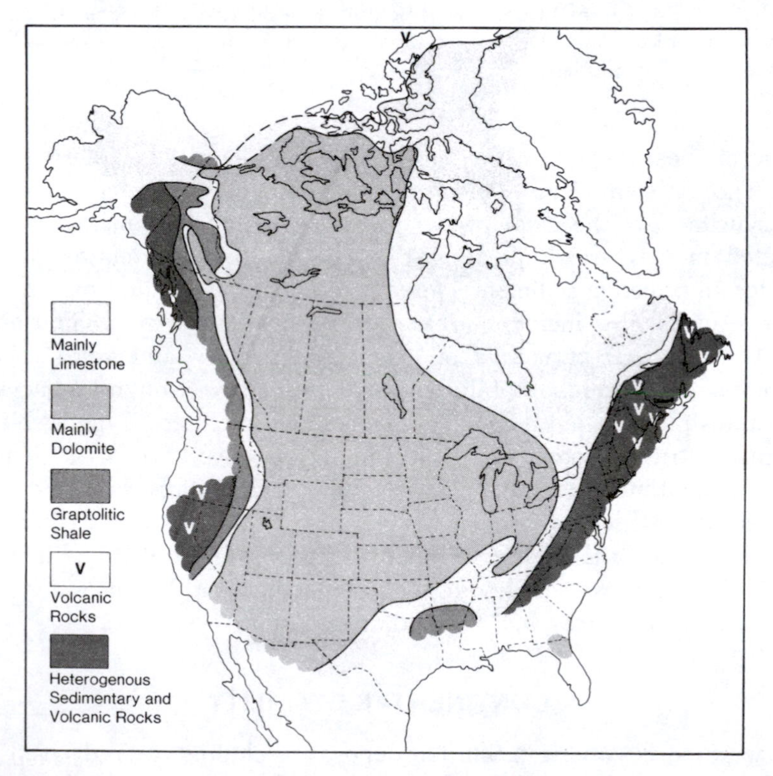

Figure 5.3. Restored Silurian lithologic map showing inferred distribution of extensive carbonate sheet covering central part of North America including the shield. Marginal geosynclines have more diverse rocks reflecting Taconic orogenic lands along the east coast as well as volcanic islands. Most of the dolomite of the continental interior was first deposited as limestone, then dolomitized.

BACK-REEF EVAPORITES IN NORTH CENTRAL UNITED STATES

The rubber, chemical, plasterboard, and photographic industries of Michigan, Ohio, and western New York owe their present location, at least in part, to a paleogeographic peculiarity that resulted in extensive Silurian salt and gypsum beds in these states. Accumulation of hundreds of metres of rock salt and gypsum is the result of the evaporation of many kilometres of seawater – not that the Silurian sea was ever deep here, but rather that shallow marine basins served as evaporating pans. These basins had limited access to the open seas and no influx of fresh water. The chief barrier between these "evaporating pans" and the open sea was a series of Silurian reefs or limestone structures built by the accumulation of the skeletons of shallow-water organisms which, as figure 5.4 shows,

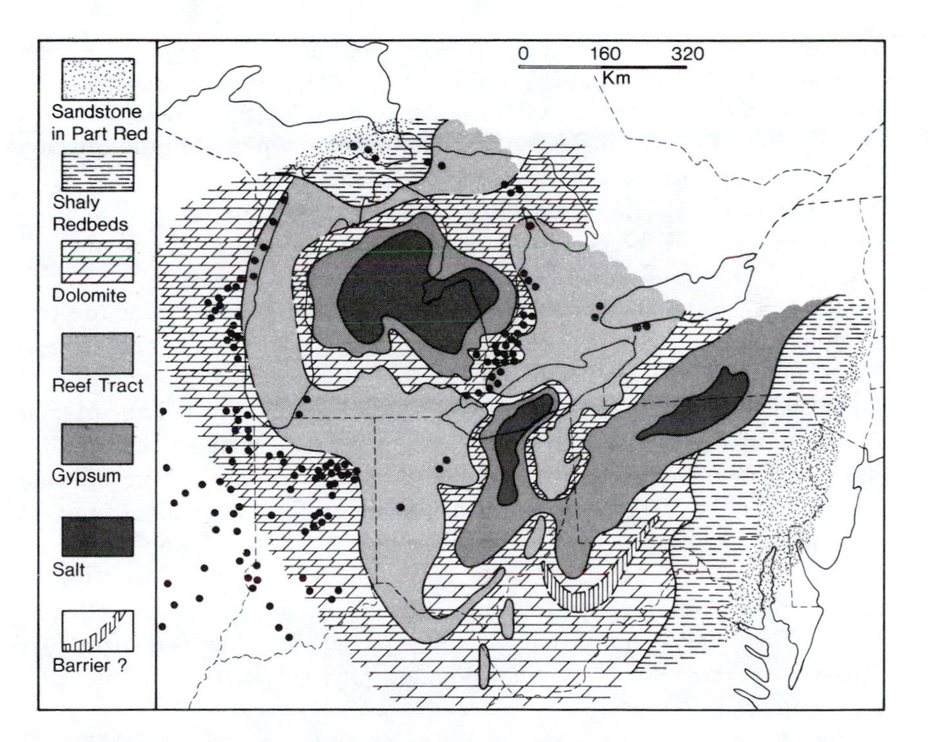

Figure 5.4. Silurian evaporite basins in eastern North America. Dots represent Middle Silurian coral and stromatoporoid patch reefs, which formed the reef margins from which the Upper Silurian reef barriers developed. Reefs served as barriers between the basins and continental seas except on the east side, where a wedge of mud and sand eroded from the Taconic land mass and formed the barrier. (After Alling, H. L., and Briggs, L. I., Am. Assoc. Petrol. Geol. Bull. v. 45, pp. 517-547, 1961)

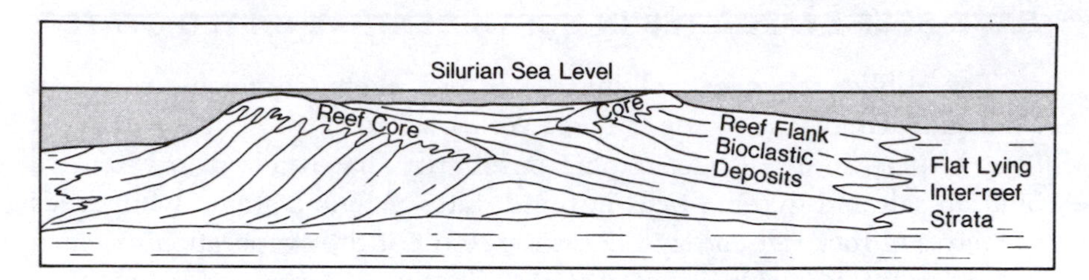

Figure 5.5. Middle Silurian patch reef complex at Thornton, Illinois, near Chicago. Wave-resistant reef core was built principally by corals and stromatoporoids and provided a base for trilobite, cephalopod, and crinoid communities. Reef-flank bioclastic deposits include reef core fragments as well as brachiopods, gastropods, and sponges. Sea water barely covered the reef core but was perhaps 100 metres deep in inter-reef areas.

virtually surrounded the area of salt accumulation. Calcareous algae have formed reefs since Precambrian time, but the Silurian marks the first time that reefs included areally extensive structures built by more complex communities. Figure 5.5 shows the structure of a typical Middle Silurian patch reef, and figure 5.4 shows locations where reefs of this age have been found. Corals and stromatoporoids built the main core of Silurian reefs; and crinoids, brachiopods, and molluscs lived in the back-reef, lagoon, and reef-flank deposits.

Silurian strata of the midcontinent are now largely dolomitic, although they are composed of accumulations of whole and broken seashells and reef materials that were undoubtedly originally composed of limestone because no known organism secretes dolomite. The original limestone was probably dolomitized very shortly after deposition as conditions of temperature, salinity, and alkalinity in the warm Silurian epicontinental seas favored the exchange of magnesium for calcium.

Middle Silurian patch reefs formed the foundation for Upper Silurian reef tract barriers between the New York, Ohio, and Michigan salt basins and the surrounding epicontinental seas. Only along the eastern margin of the New York-Pennsylvania basin did red-shales eroded from the Taconic uplift serve as another type of physical barrier limiting the evaporite basin.

MICHIGAN AND WILLISTON BASINS

Two areas stand out on figures 5.1 and 5.6 as having accumulated unusually thick Silurian deposits, considering their location on the craton: the Williston Basin in North Dakota, and the Michigan Basin. Both are filled with carbonates and evaporites, and their style of accumulation

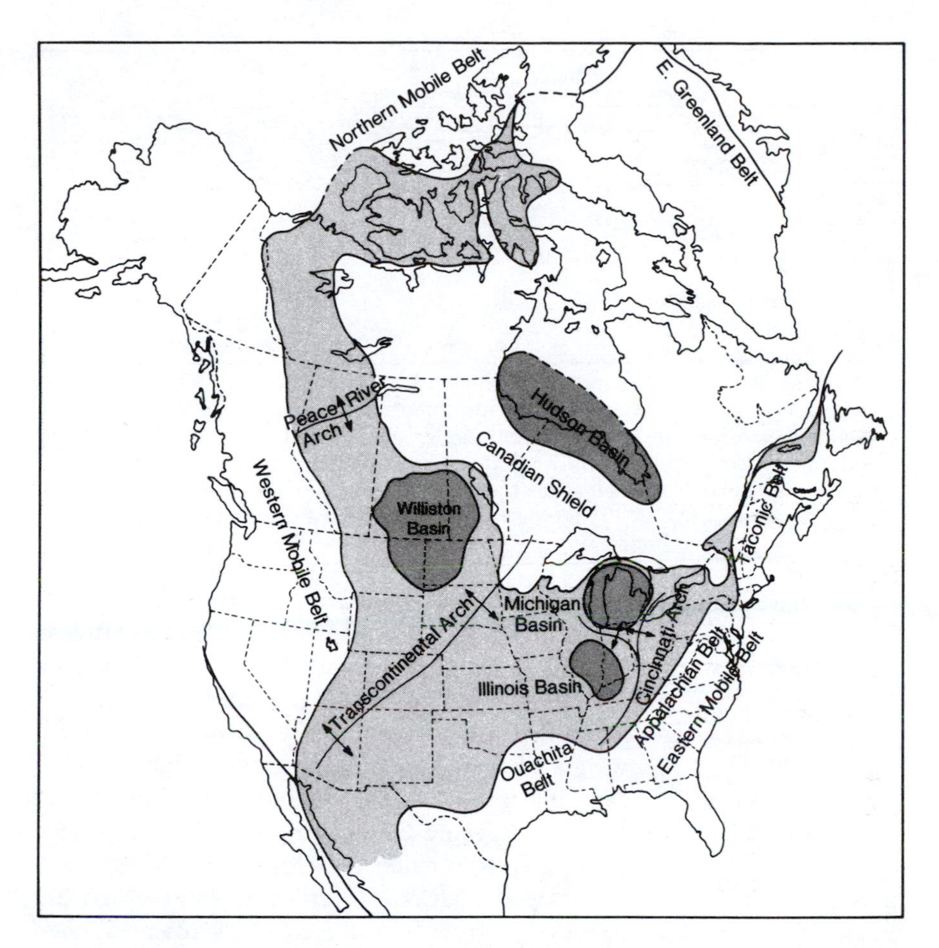

Figure 5.6. Map showing major Paleozoic basins and uplifts of North America.

tells us something about the rigidity of the earth's crust and the localization of thick piles of sediments.

The region of Michigan and parts of the adjacent states and Canada is referred to by geologists as the Michigan Basin not only because of the great circular downwarp, which centers on the state, but also because it has been found that strata, particularly the Silurian, which are exposed around the edges of the Michigan Basin, become substantially thicker in the center of the downwarp, as shown in figure 5.7, where penetrated by drilling operations. Thus this basin was subsiding more rapidly in its central part in Upper Silurian time than it was around its edges.

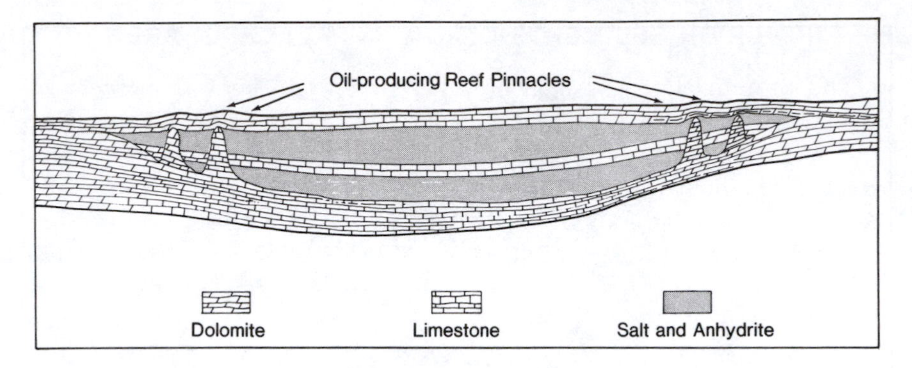

Figure 5.7. Cross section of the Michigan Basin showing Silurian lithologic relationships as known from oil well records. Reef pinnacles on either side of the basin serve as oil traps. The oil probably originated in the basal limestone, a laminated algal bed, and migrated into the pinnacles. Note increase in thickness of evaporite in the center of the basin.

We have observed that any load on the earth's crust will cause temporary subsidence: Pleistocene glaciers 3 kilometres thick in the Hudson Bay area temporarily depressed the crust up to 100 metres, as indicated by peripheral ice age shorelines that are now inclined away from the center of greatest depression because of elastic rebound following removal of the ice load. The weight of the water in Lake Mead behind Hoover Dam – to give a modern illustration – has depressed the underlying crust about 15 centimetres. The Taconic Orogeny resulted in a great pile of clastic sediments that apparently pressed the earth's crust down as it accumulated. The abnormal accumulation of carbonates and evaporites in the Michigan Basin, however, differs from the crustal loads mentioned above in that it did not result from some external loading mechanism but rather was the result of self-contained, slow subsidence of the basin, the origin of which might represent heat transfer movement and softening in the mantle beneath the crust at this spot. Some have said that such intracratonic basins were caused by "inverted mantle plumes." In any event, the Michigan Basin resulted from the remarkably close adjustment between basin subsidence, reef building, and salt accumulation. A total of almost a kilometre of Late Silurian evaporites, all deposited at sea level, are found in the center of the Michigan Basin. The Williston Basin includes a similar lens-shaped chemical deposit, chiefly dolomite, about 300 metres thick at its center.

IRON-RICH SEDIMENTS OF THE APPALACHIAN AREA

The Birmingham, Alabama, area has produced about 10 percent of the nation's iron from Silurian sedimentary rocks known as "Clinton iron ores" named from their occurrence near Clinton, New York. The ores extend throughout the Appalachian region but are of commercial thickness only in the south. Facies relations of the hematite ores indicate that they were deposited in a lagoonal environment. The eroded Taconic Mountains were situated along the eastern seaboard and barrier islands paralleled their western coastline creating lagoons between the islands and the eastern land mass. Within the central parts of the lagoons, iron minerals were reduced, whereas along the marginal parts oxidized hematitic sandstone and iron ores were developed. Iron was supplied by streams flowing into the lagoons from the eastern land mass. Chemistry of such iron transportation and deposition suggests that the source area consisted of metamorphic and igneous terrain, which was humid, poorly drained, and deeply weathered. Broken brachiopod shells that occur in some of the ore beds were probably transported eastward from their place of growth on the shallow water shelf of the midcontinent epicontinental sea floor, across the nearshore barrier islands during storms and deposited in the quiet lagoonal waters, where they became replaced by iron oxide that was brought into the lagoon by streams eroding the eastern land mass.

SILURIAN LIFE

Sir Roderick Murchison realized the critical role that fossils play in evaluating the age of strata. When he published his account of the *Silurian System* in 1839, he included not only measured stratigraphic sections, but also described and illustrated Silurian fossils. The type British Silurian is characterized by a "shelly" fauna of corals and brachiopods in carbonate rocks, and by graptolites in other parts of Great Britain where the rocks are shaly. Thus, just as in the Ordovician, there are two major faunal facies groups: graptolite-bearing shales and shell-bearing carbonates. These same faunal facies relationships exist in Silurian rocks in many parts of the world. Within Silurian strata in North America carbonates are much more prevalent than shales; therefore, brachiopods and corals are the most common Silurian fossils. Solitary horn corals, chain corals, and honeycomb corals, shown on figure 5.8, are common in Upper Ordovician strata and abundant enough in Silurian time to form part of these early coral reefs.

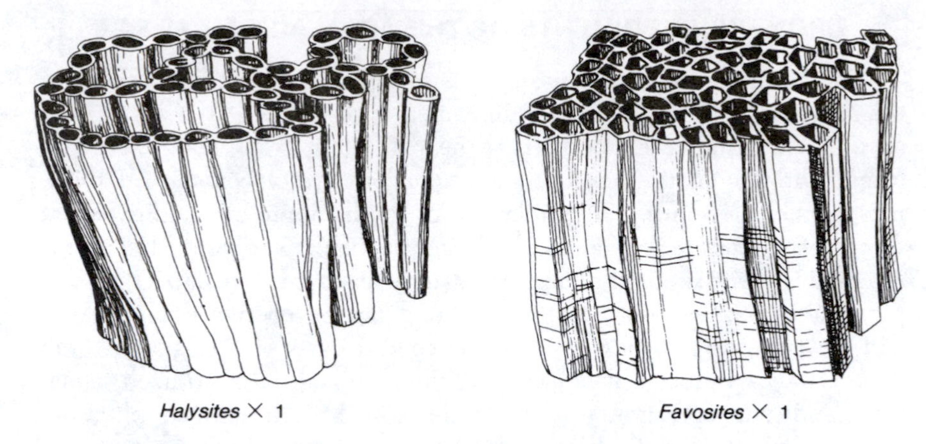

Figure 5.8. Typical colonial corals found in Silurian limestones and dolomites in North America. **Halysites** is called the "chain" coral. **Favosites** is known as a "honeycomb" coral. Both are common in Late Ordovician and Silurian carbonates.

Graptolites evolved through many genera with a considerable range in size and shape during the Ordovician. They are less diverse in the Silurian, most being characterized by having the living chambers arranged vertically in a row on one side of the connecting column. This form is called monograptid after the typical genus *Monograptus* (fig. 5.9). In the shaly facies where they occur they are excellent zone fossils with short-lived species of widespread geographic distribution. Twenty-nine monograptid graptolite zones, shown on figure 5.10, are recognized in Great Britain and Germany; many of the same British Silurian graptolite zones have been identified in North American shaly sequences, as the right-hand column in figure 5.10 indicates. This British graptolite zonation is so refined that it permits us to recognize Silurian time increments on the order of a million years in length, no mean feat for so distant a time!

In carbonate rocks, Silurian shelly faunas include, in addition to corals and brachiopods, trilobites, crinoids, cephalopods, conodonts, gastropods, and sponges. The predominant impression of Silurian shelly faunas is of cosmopolitan distribution, perhaps reflecting the circumstance that most observable Silurian rocks lie along the trend of the Silurian equatorial belt. The Iapetus Ocean, which may have been wide enough to serve as a barrier to migration of provincial organisms in Cambrian and Ordovician times, apparently became closed during Silur-

Figure 5.9. Three different kinds of monograptid graptolites. Monograptids are typified by having the living chambers arranged in a single vertical row along a hollow tube. Monograptids dominate Silurian graptolite assemblages.

ian time as indicated on figure 5.1. Continental platforms were exceptionally broad during the Silurian, thus permitting wide distribution of uniform faunas. Another factor that promoted cosmopolitan distribution of Silurian faunas was the Late Ordovician-earliest Silurian continental glaciation of Africa and South America, which took water out of the oceans and caused restriction of the Late Ordovician epicontinental seas. Melting of these glaciers in early Silurian time caused the sea level to rise and permitted wide dispersal of Silurian marine faunas.

Ostracodes, small bivalved crustaceans, are abundant and useful guide fossils wherever found, being limited mostly to occurrence in shales and shaly limestones. Another interesting arthropod that flourished during the Silurian is the eurypterid or "sea scorpion." Commonly 10-20 centimetres and up to 1.8 metres in length, bearing long clasping appendages, they were logical competitors for the trilobites and may have accounted

PERIOD	EUROPEAN SERIES	BRITISH ISLES	CANADA
DEV.			
SILURIAN	PRIDOLIAN		Monograptus sp. T
			Monograptus sp. P
			Monograptus ultimus
	LUDLOWIAN	29. Bohemograptus bohemicus	
		28. Saetograptus leintwardinensis	
		27. Pristiograptus tumescens	= P. tumescens minor
		26. Lobograptus scanicus	Monograptus sp. 0
			Monograptus bohemicus
		25. Neodiversograptus nilssoni	= N. nilssoni
	WENLOCKIAN	24. Monograptus ludensis	Monograptus testis
		23. Gothograptus nassa	
		22. Cyrtograptus lundgreni	= C. lundgreni
		21. Cyrtograptus ellesae	Cyrtograptus perneri
		20. Cyrtograptus linnarssoni	C. sp. G, C. sp. F
		19. Cyrtograptus rigidus	= C. rigidus
		18. Monograptus riccartonensis	= M. riccartonensis
		17. Cyrtograptus murchisoni	= C. murchisoni
	LLANDOVERYAN	16. Cyrtograptus centrifugus	Stomatograptus grandis
		15. Monoclimacis crenulata	Monograptus spiralis
		14. Monoclimacis griestoniensis	
		13. Monograptus crispus	
		12. Monograptus turriculatus	= M. turriculatus
		11. Rastrites maximus	
		10. Monograptus sedgwickii	
		9. Monograptus convolutus	Monograptus millepeda
		8. Monograptus argenteus	
		7. Diplograptus magnus	
		6. Monograptus triangulatus	
		5. Coronograptus cyphus	= C. cyphus
		4. Lagarograptus acinaces	
		3. Atavograptus atavus	
		2. Orthograptus acuminatus	
		1. Glyptograptus persculptus	
ORD.			

Figure 5.10. Silurian graptolite zones of the British Isles reference section compared to zones found in North Canada. This diagram illustrates the degree of time resolution allowed by graptolite zonation for the Silurian Period. These zones permit intercontinental correlation of Silurian rocks and events. (After Berry, W. B. N., and Boucot, A. J., 1970, Geol. Soc. Am. Special Paper 102, 289 p., 1970, and Rickards, R. B., 1976, Geol. Jour., v. 11, pp. 153-188)

for the striking decline of the latter after the Silurian. The Silurian Period has been called the "Age of Eurypterids" after these invertebrates, although their importance is one of uniqueness more than abundance.

For the first time in the geologic record plants were present on Silurian continents. Marine plants are some of the oldest known forms of life on earth, yet prior to the Late Silurian, land plants did not exist. Found in eastern Australia in cherts bearing typical Late Silurian graptolites, the plant fossils contain a well-preserved vascular bundle, a necessary adaptation for plants to exist on land. These Australian specimens represent the remains of land plants washed into the sea and deposited with marine sediments.

SILURIAN TIME SUBDIVISIONS

The Silurian System serves as a good example of how geologic time subdivisions are originally established and later modified. Roderick Murchison named the Silurian in 1839, describing not only the rocks of that age in Wales but also their contained fossils. The lower limit of the British Silurian is at the base of shelly fossil-bearing Llandovery sandstones, and the subdivisions of the type section as recognized by Murchison are in ascending order: Llandovery, Wenlock, and Ludlow. Llandovery and Wenlock age rocks are represented in the British Isles by both shelly and graptolite faunal sequences. The Ludlow contains only graptolites. Since Murchison's time many paleontologists have studied Silurian-age fossils collected from many parts of the world and have come to realize that the British fossil sequence is useful as a worldwide correlation standard. Furthermore, the British graptolite zones themselves have been intensively restudied over the years reaching the level of refinement shown on figure 5.10 only recently. The upper Ludlow in Great Britain does not contain fossils useful for long-distance correlation; instead, the Czechoslovakian Silurian-Devonian sequence area has been utilized to paleontologically define the Silurian-Devonian boundary. The Czechoslovakian sequence is more continuously fossiliferous with graptolites and shelly faunas than is the British Upper Silurian. Therefore, it is used as the world standard of reference using Pridoli beds for the uppermost Silurian subdivision. Thus refinement of our knowledge of distribution of organisms in time and space ultimately results in a network of paleontologic information, which we interpret to determine relative time.

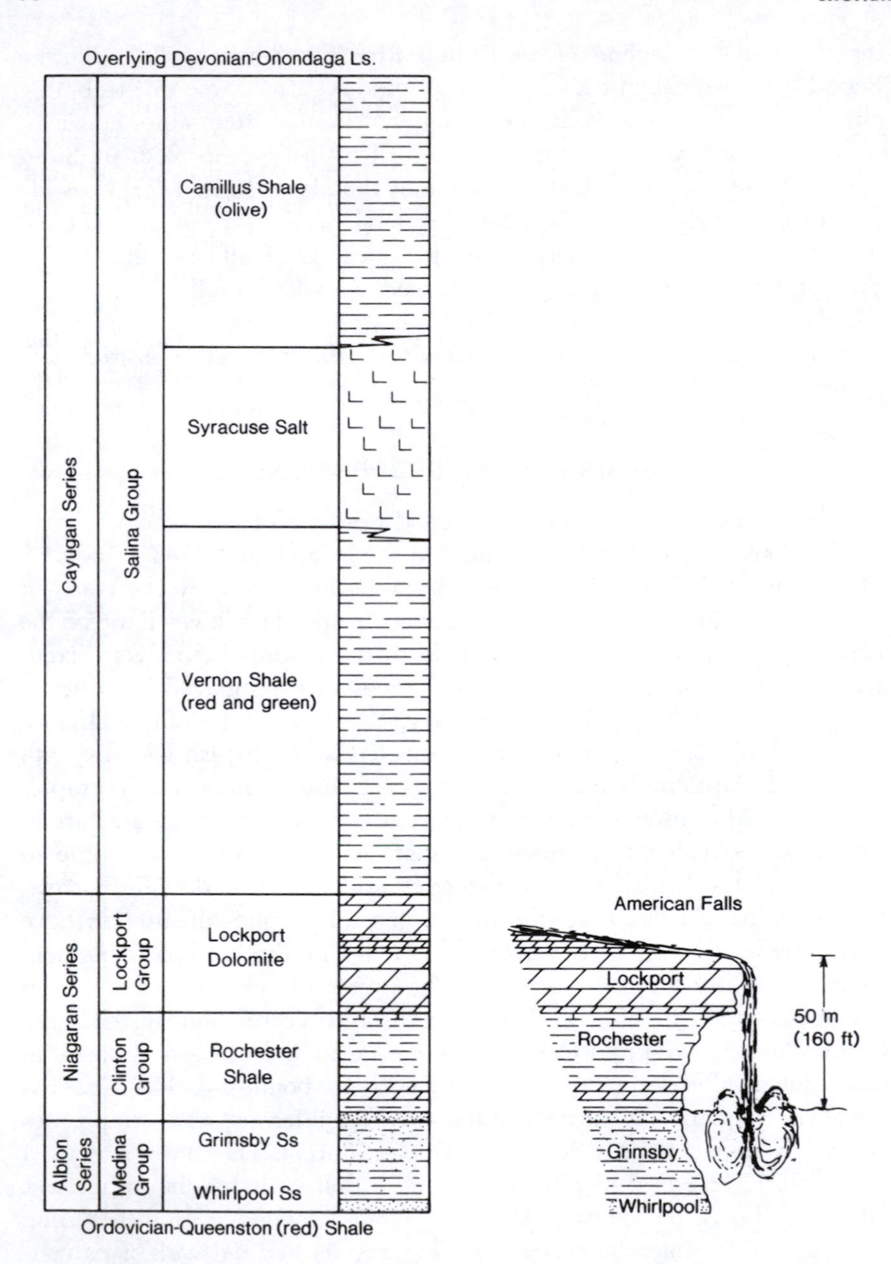

Figure 5.11. Silurian strata in eastern New York. Middle (Niagaran) and Lower (Albion) Silurian strata are present at Niagara Falls. The less resistant Upper Silurian (Cayugan) strata form nearby low topography largely covered with Pleistocene glacial deposits. The Syracuse Salt is not exposed at the surface but is encountered in wells to the east near Rochester.

NIAGARA FALLS SECTION

The area around Niagara Falls has spectacular exposures of Silurian rocks that extend across New York state and encircle the Great Lakes. New York Silurian stratigraphy, shown on figure 5.11, was studied and defined more than 100 years ago, and established about 1900 as the North American Silurian standard section. As a standard, unfortunately, it has not proved to be useful in other parts of North America. The New York Silurian is the product of unusual environments of deposition, discussed earlier in this chapter, and most of the fossils contained in the section are not widely distributed elsewhere in North America. The Lower Silurian (Albion Series) is relatively unfossiliferous sandstone; the middle subdivision (Niagaran Series) is the most fossiliferous part of the New York section and has been used for correlation, but the upper subdivision (Cayugan Series) consists of beds deposited in shallow waters of high salt content containing few fossils. Experts in Silurian fossils have suggested that the New York section be abandoned as a North American standard, since more consistent comparisons can be made directly with the established European Silurian sequence. Figure 5.12 compares the two systems of subdivision.

British Divisions		North American Divisions
Pridolian (Skalan)		Cayugan
Ludlowian		Niagaran
Wenlockian		
Llandoveryan	Upper	
	Middle	Albion (Medinan)
	Lower	

Figure 5.12. Silurian subdivisions. British subdivisions are becoming increasingly used in North America because the British strata are more continuously fossiliferous than the North American subdivisions. The latter were established in eastern North America in rocks representing environments of limited areal distribution.

SILURIAN TECTONICS

Except for the residual effects of the Ordovician Taconic Orogeny, expressed as Early Silurian sandstones and shales in the Appalachian area, eastern North America was in a period of tectonic quiescence during Silurian time. The Caledonian Orogeny that affected northwestern Europe and eastern Greenland during the Silurian (see fig. 5.1) did not extend into North America. Closing of the Iapetus Ocean brought Europe and North America into their closest proximity; Africa (and Florida) was yet to arrive off our eastern seaboard.

Andesites and rhyolites, shown on figure 5.3 by Vs, are reported from several localities between Maine and Newfoundland. These scattered occurrences of Silurian volcanic rocks are interpreted to mean that a subduction zone and associated volcanoes were active along the northeastern margin of the continent.

Along the west coast of North America the Silurian is represented by a variety of clastic rocks, including graywackes interbedded with coral and brachiopod-bearing limestone lenses. Volcanic rocks are less common than on the east coast, but include flows and intrusions dated radiometrically as early Silurian. Whether these igneous rocks were formed as part of the Silurian North American continent or whether they were formed elsewhere in an offshore island arc that became later attached to western North America has not been determined because of the complexity of later tectonic events in the area. In any event, these volcanic rocks bespeak a Silurian subduction complex that is now a part of western North America.

6

Devonian

TOPICS

History of Studies
Devonian Transgression
Middle and Late Devonian Reefs
Devonian Life—Age of Fishes
Devonian Mountain Building
Devonian Tectonics

HISTORY OF STUDIES

In 1837 W. Lonsdale, an Englishman, described fossil corals from marine rocks in Devonshire in southwestern England and concluded they were intermediate between those from underlying Silurian rocks and those from the overlying lower Carboniferous rocks. Two other Englishmen, R. I. Murchison and Adam Sedgwick, proposed the Devonian System for these intermediate strata in 1839, based in part upon Lonsdale's coral data. Approximately 50 million years of earth history are recorded by Devonian rocks; they document ancient mountain-building movements and widespread coal reef environments, as well as major diversification in the evolution of fish, land plants, and invertebrate animals during the period.

The type locality for the Devonian system is in Devonshire, but the more fossiliferous strata in the Rhine Valley of Belgium and Germany have become the standard for Devonian studies throughout the world. The abundantly fossiliferous Devonian rocks in New York are used as a standard reference section for North America in conjunction with those exposed along the Rhine River Valley. Figure 6.1 illustrates the distribution of Devonian rocks in North America.

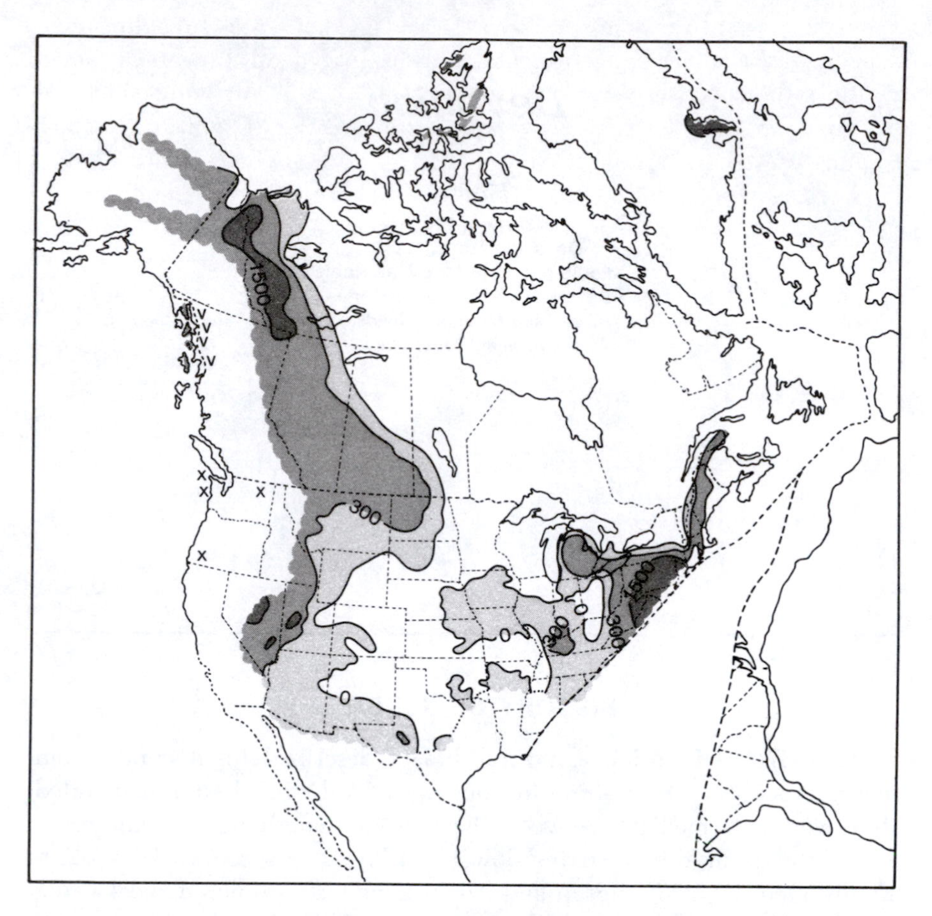

Figure 6.1. Thickness map of Devonian System in North America. Thicknesses shown in metres. Thickest sections reflect Late Devonian mountain building around the margin of the continent.

DEVONIAN TRANSGRESSION

Lower Devonian rocks are absent throughout much of the interior of North America and a widespread unconformity generally separates Middle Devonian and Middle Silurian strata, as a response to almost complete withdrawal of seas from the interior of the continent (fig. 4.7). The continental interior of North America was undergoing erosion during most of Early Devonian time. The most nearly complete sequences of

Lower Devonian rocks are those preserved in the gently subsiding geosynclinal areas along the continental margin. In general, Lower Devonian clastic sediments were derived from erosion of the low-lying shield of the continent and sediments were spread toward the margin in a pattern reminiscent of Cambrian, Ordovician, and Silurian beds. This pattern changed, however, during the Middle Devonian so that for much of the remainder of Paleozoic time, sediments were shed from uplifts within the geosynclinal areas and onto the continent margins.

Middle Devonian seas spread across the eroded surface of the interior of the continent and deposited limestone, shale, and some sandstone. These seas spread from the Appalachian geosyncline westward to the Mississippi valley to meet a seaway that had extended southward along the subsiding western margin of the continent across western Canada. These also joined a seaway that had extended eastward from the Cordilleran trough in the western United States. This broad expanse of warm, shallow marine water on the continent provided extensive new and favorable environments for both vertebrates and invertebrates, judging from the abundant fossils in rocks deposited in the seas.

Some of the clean, quartzose Devonian sandstones are thought to have resulted from reworking of mature soils during the transgression of the seas. The Oriskany Sandstone (fig. 6.2) is a good example of one of these sheets of sandstone and is excellently exposed in New York and Pennsylvania, where it marks the base of a transgressive Lower Devonian pulse.

Cyclic transgression and regression of seas onto continents on a worldwide scale are thought by many geologists to represent not movements of single continents but some broader phenomena. Lowered sea level during some periods is thought to represent withdrawal of water from the sea during times of major glaciation. However, this is not true for the Devonian, for not all times of regression of the sea are marked by glacial events. Other scientists have concluded that the mechanism of transgression and regression may be linked to motion and interaction of major plates. Transgressions are, in general, associated with times of rapid plate motion, concurrent with major upwarps along midoceanic spreading centers. Thus water is crowded from the ocean basins and onto the continents. Conversely, regressions may be associated with periods of worldwide minimal movement of plates when the floors of the ocean basins were not so highly warped. Those times of major

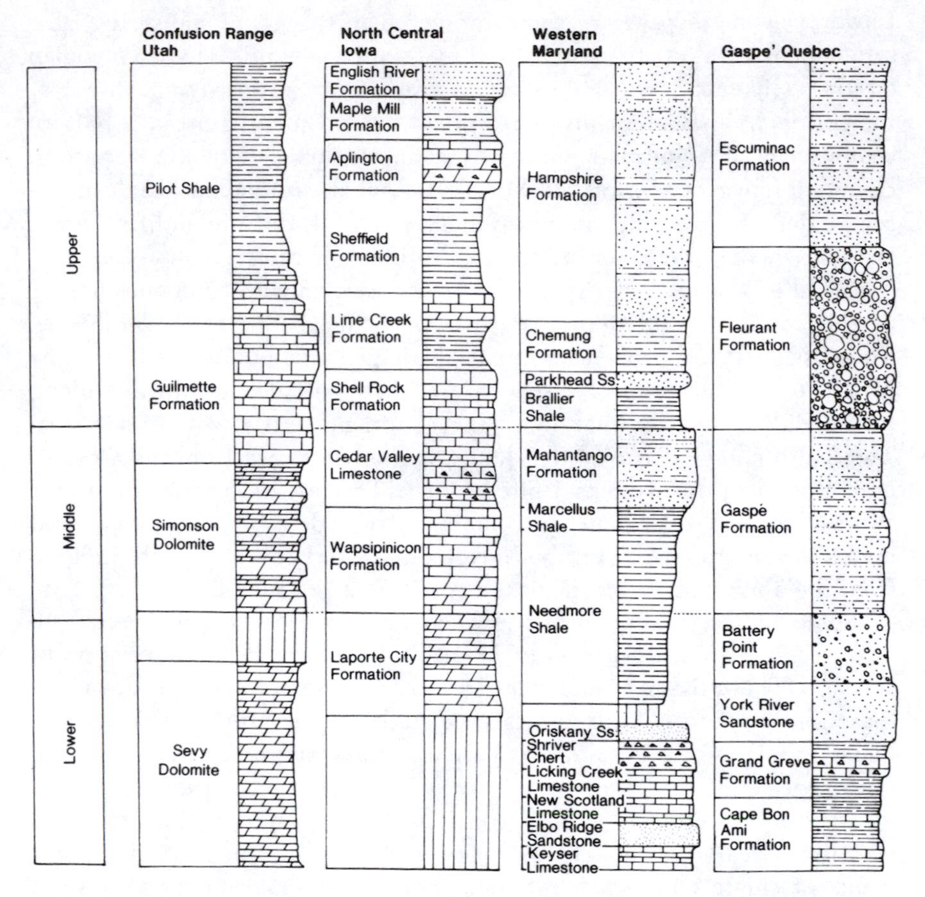

Figure 6.2. Representative stratigraphic sections of Devonian rocks throughout North America. From hundreds of similar sections geologists gather data to interpret the geologic history of an area. Names are formation designations. The vertical scale indicates the time span, not thickness of the strata, in a manner similar to figure 3.7D. Vertically ruled areas indicate nondeposition.

withdrawal of shallow seas from the continents, such as in Early Cambrian, Early Devonian, and Early Mississippian (fig. 4.7) may mark times of particularly reduced plate motion.

The latest Devonian and earliest Mississippian are marked by one of the most widespread black shale sections in the geologic record in North America. The Chattanooga Shale and equivalent units were deposited on the southeastern and eastern part of the shelf at the same time that the Exshaw Shale of the Canadian Rocky Mountains and equivalent beds in the plains to the east were deposited. Contemporaneous black shale covers much of the Williston Basin and appears to be part of a sheet of gray shale that spreads over much of Utah and Nevada as well. The Chattanooga Shale rests on a remarkably even erosional surface and in Tennessee alone rests on twenty-three different sedimentary formations that range from Middle Ordovician to Middle Devonian age. Some scientists have suggested that these remarkably widespread, very thin black shales represent fossil soils. Most of the formations, however, contain marine fossils in addition to terrestrial plant debris. Other workers have suggested that the shales accumulated in unusual deep basins on the craton. A more recent model suggests that these black shales may have accumulated in shallow water beneath a floating cover of extensive rafts of seaweeds that protected the water and the sediments on the bottom from aeration. Debris from the decomposing lower parts of the raft would have further depleted any available oxygen such that life would have been impossible for normal scavanging bottom dwellers.

MIDDLE AND LATE DEVONIAN REEFS

Devonian reefs, resembling those of Silurian age in the Michigan Basin in the Great Lakes area, formed in western and southern Alberta (fig. 6.3) and influenced sedimentation along much of the western shelf of Canada and the north central part of the United States. These reefs are composed mainly of skeletal remains of stromatoporoids, corals, and calcareous algae. The reef structures are massive and obscurely bedded (fig. 6.4) and are now highly porous because of alteration of the original calcareous skeletal material to dolomite. As a consequence these reefs are excellent reservoirs for petroleum.

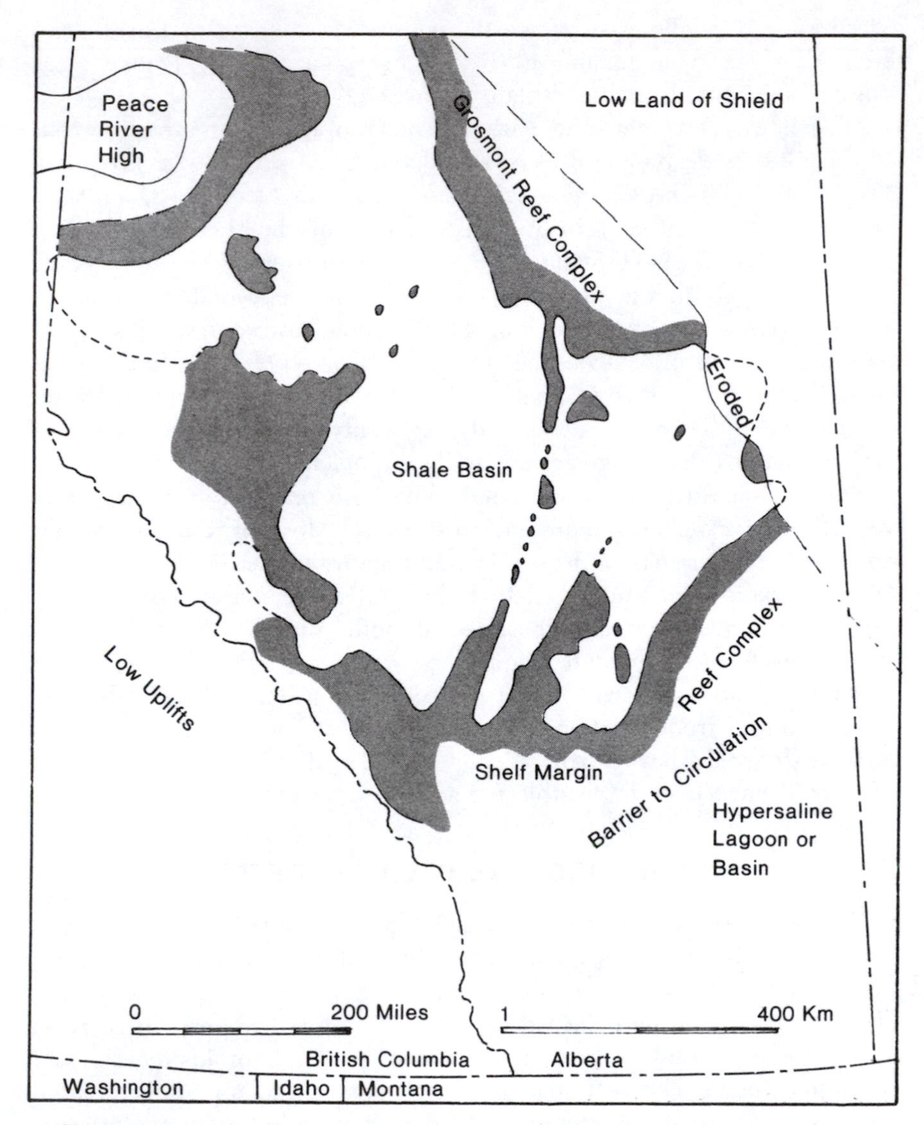

Figure 6.3. Map of southern Alberta and British Columbia showing Devonian reef distribution, marginal land areas, and the inner reef shale basin. Connection to the open sea was to the northwest beyond the Peace River High.

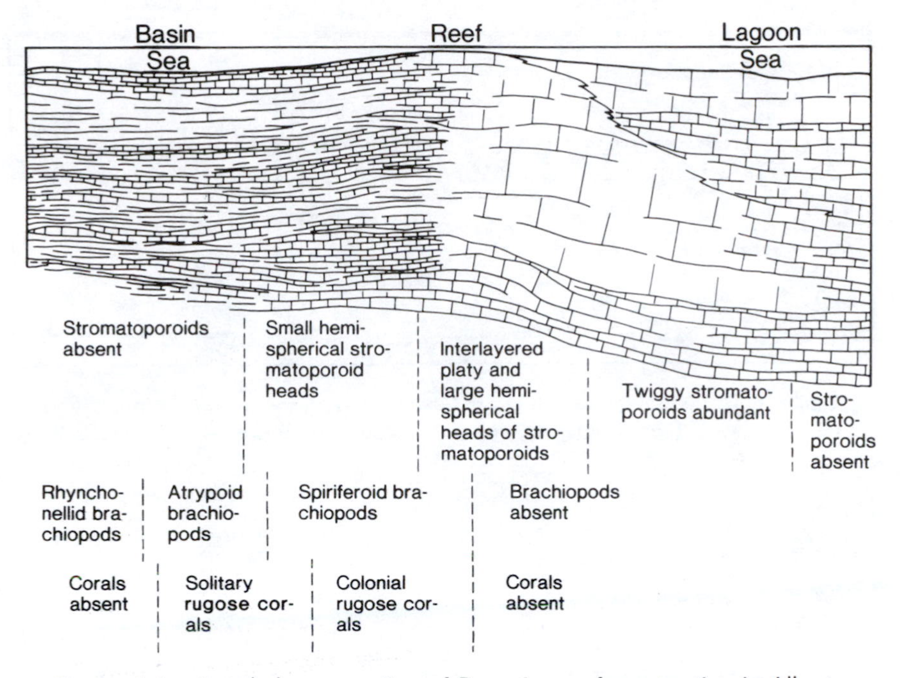

Figure 6.4. Detailed cross section of Devonian reef construction in Alberta. Oil is recovered from the porous limestones and dolomites of similar massive reef buildups.

Thse reefs flank the margin of the shield, to the northeast of the basin, and developed around several uplifts to the northwest and southwest of the basin (fig. 6.5). To the southeast, however, they produced a barrier to free circulation of sea water and separated the normal open marine conditions of Alberta and British Columbia from the hypersaline evaporite basin of part of the Williston Basin. Salinity reached its peak within the lagoon in Middle Devonian when salt, anhydrite, and potash accumulated up to 300 metres thick in the Williston Basin. Salt deposits rarely crop out because they are so soluble, but they are well known in the subsurface by drilling and mining. Evaporite deposition ceased in the Upper Devonian when more nearly normal marine water spread across the shelf area and over the barrier.

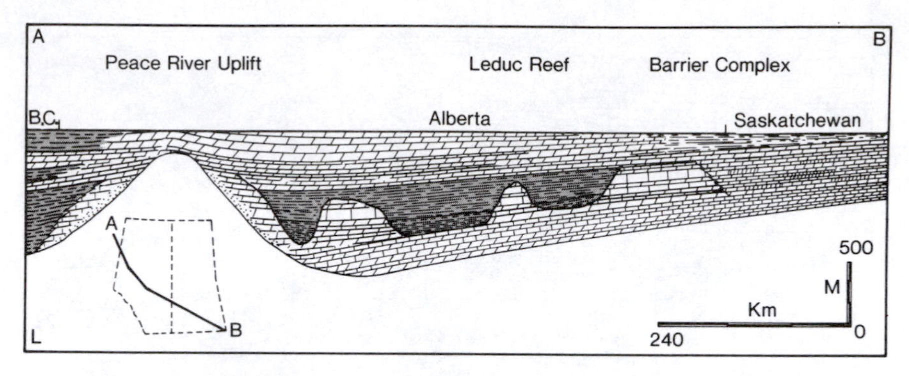

Figure 6.5. Cross section through Devonian reef complex in Alberta and Saskatchewan. These reefs are major oil producers in western Canada.

Figure 6.6. Ammonoid, showing characteristic simple suture pattern of Paleozoic forms. Rapid modification of the suture allows these fossils to be utilized for recognition of fine subdivisions of the stratiographic sequence, even continents apart.

DEVONIAN LIFE—AGE OF FISHES

Devonian strata contain many exciting new occurrences of both plants and animals. Ammonoid cephalopods (fig. 6.6), whose shells serve as indexes for intercontinental correlation for Upper Paleozoic and Mesozoic rocks, first appeared in the Devonian. Stromatoporoids, now considered to be relatives of sponges, colonial and solitary corals, and algae combined to form reef tracts not only in western Canada but at many places in western North America. Conodonts and spiriferid brachiopods (fig. 6.7) flourished in Devonian seas as well, judging from their abundance as fossils.

Land plants, although first known from the Silurian, became abundant and formed forests for the first time, and from the Devonian onward the continents were at least partially covered with vegetation.

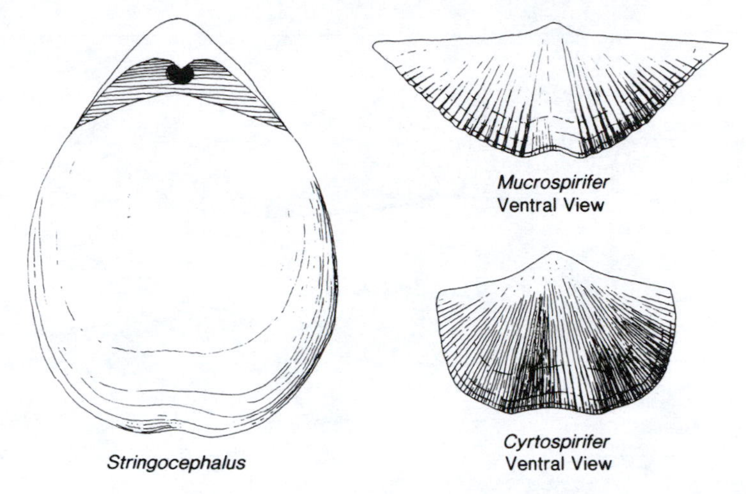

Figure 6.7. Common brachiopods of the Devonian. **Stringocephalus** is a characteristic Middle Devonian brachiopod. Spiriferid forms are especially abundant in rocks of this age.

Devonian plants were small, however, for most were no more than one metre high. A few fossil trees as high as 10 to 15 metres are known from Upper Devonian strata in New York.

In spite of significant developments in plants and invertebrates, the Devonian is known as the "age of fishes" largely because these are the oldest rocks where fossil fish are moderately abundant. Armored placoderms (fig. 6.8) evolved from the jawless ostracoderms that were the dominant fishes of the Ordovician and Silurian. One group, the arthrodires, were as much as 10 metres long and were certainly the most impressive carnivores of their time. Primitive sharks, bony fish, lung fish, and lobe-finned crossopterygian fish all arose from the placoderms during the Devonian and are known in both marine and fresh-water deposits. Amphibians, the first vertebrates to venture onto the land, developed from the crossopterygian fish in the Devonian and occur for the first time in Upper Devonian rocks of eastern Greenland.

DEVONIAN MOUNTAIN BUILDING

By the close of the Devonian period, three major mountain ranges had been formed around the continental margins of North America (fig. 6.9). The Antler Orogeny, well documented in central Nevada, produced mountains that extended north and south along the transition area be-

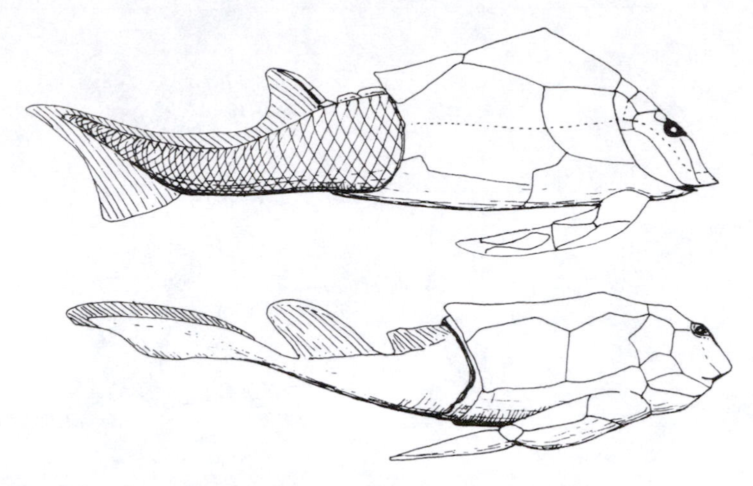

Figure 6.8. Devonian placoderms (heavily armored fish) which replaced the jawless ostracoderms. **Pterichthyodes** (above) approximately 12 centimetres long, and **Bothryolepis** (below) approximately 20 centimetres long. Both inhabited Devonian freshwater lakes.

tween earlier eugeosynclinal and miogeosynclinal belts and marked a change in sedimentary patterns in western North America for the remainder of geologic time. The Antler Orogeny started late in the period, for only Upper Devonian beds show significant clastic content (fig. 6.2) in Nevada and western Utah. The uplift became more extensive during Mississippian time, when even coarser clastic sediments were shed eastward into the miogeosyncline.

Prior to the Devonian, clastic units in the miogeosyncline had been derived largely from the interior of the continent, but from the Late Devonian on sediments in the miogeosyncline had mixed eastern and western sources. At times the sources were dominantly from the west. These marginal geosynclines, in which sediments were swept toward the craton, have been termed exogeosynclines, but for simplicity we will continue to use the term miogeosyncline for these nonvolcanic thick clastic wedges.

The northern margin of the continent, from Alaska to Ellesmere Island, was also involved in uplift during the Ellesmerian Orogeny of Middle and Late Devonian.

The eastern margin of the continent was involved in the Acadian Orogeny, the most intense of the three Devonian orogenic events, for rocks were folded and metamorphosed from Newfoundland through

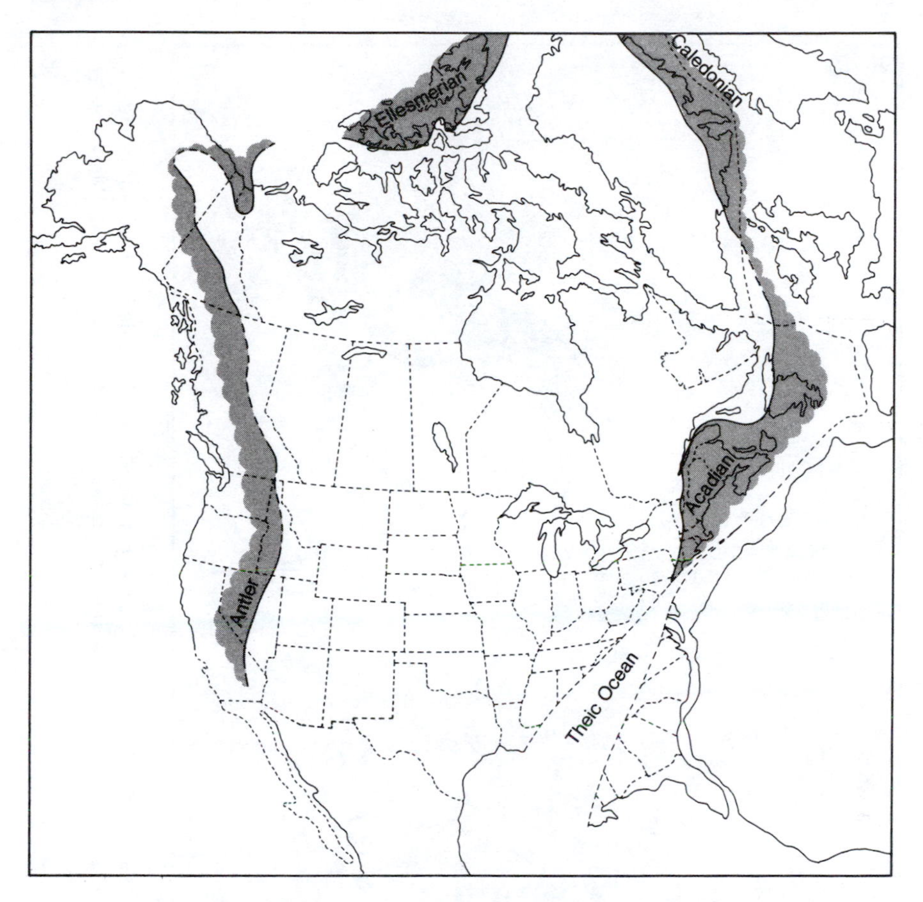

Figure 6.9. Devonian orogenic belts of North America.

New England into the Piedmont area of the Appalachian Mountains (fig. 6.10). Acadian structures are superimposed on the eroded roots of the Taconic structures. The uplift produced nearly three times as much clastic debris in eastern North America as did the Ordovician Taconic Orogeny. The Acadian event is one of the most intense orogenies to affect North America. The eugeosynclinal belt from South Carolina to the northern shores of Newfoundland was again folded and faulted. The present-day Appalachian piedmont and the metamorphosed rocks of New England reflect this repeated deformation.

The Acadian Orogeny is related to final closing of the Iapetus Ocean. Initial closure started in the north and produced the Silurian Caledonian belt of Greenland and northern Europe, and finally terminated as the

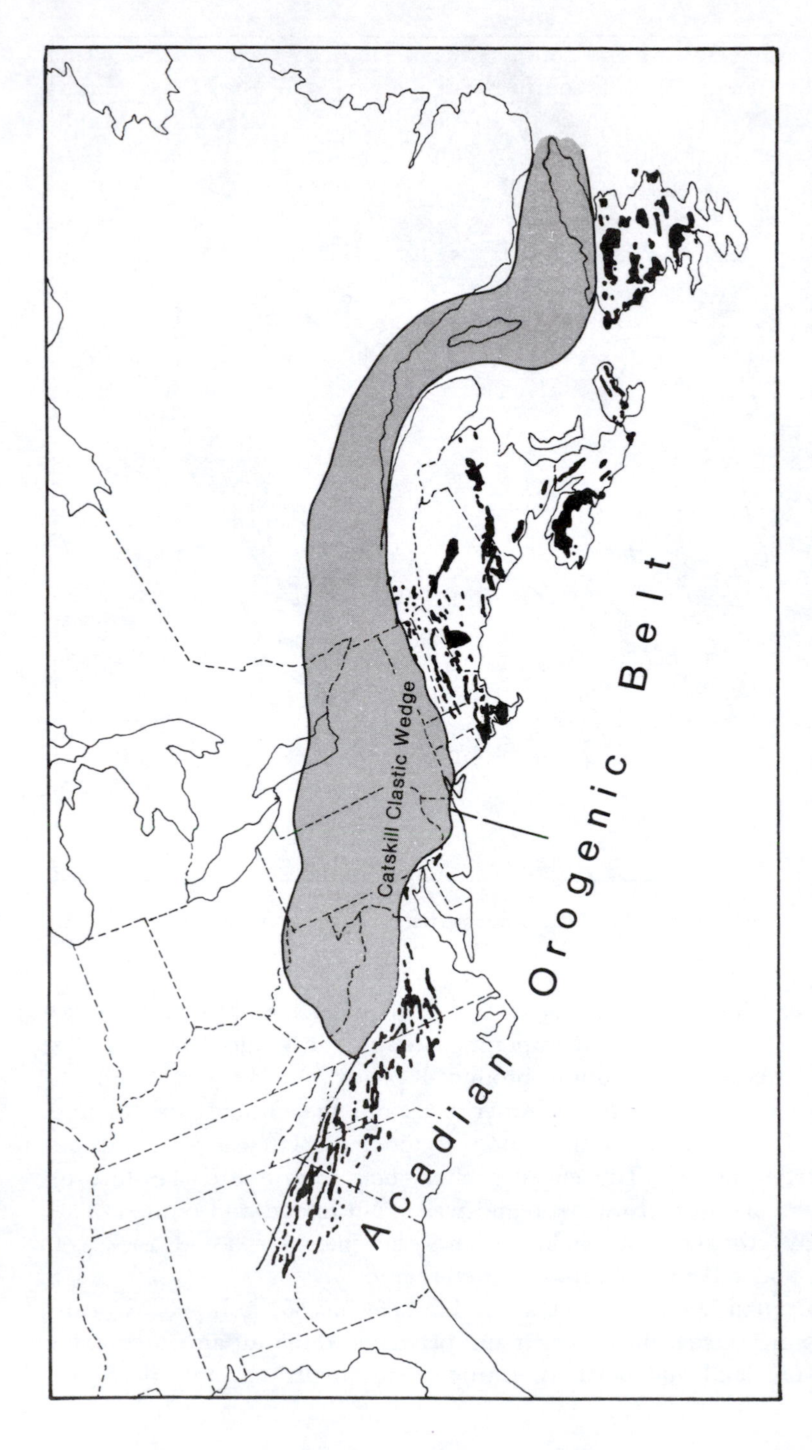

Figure 6.10. Map of eastern North America showing Acadian orogenic area (gray) and Late Devonian granite intrusions (black) which yield radiometric ages of 350 ± million years. Catskill sediments derived from the Acadian Mountains accumulated adjacent to uplift.

Acadian event toward the south. Unconformities between Lower and Middle Devonian rocks in southeastern Canada date the initial folding of the Acadian Orogeny there at approximately the base of the Middle Devonian. Coarse clastic units in Maritime Canada (fig. 6.2) show some uplift and erosion during Lower Devonian time. Lower Devonian rocks in New York and Pennsylvania show little evidence of the impending uplift, but the clastic wedge developed there in Middle Devonian time, as it did in Maryland (fig. 6.2) and areas to the southwest.

Clastic wedges developed in the marginal subsiding troughs. The wedge to the southeast is included in the Old Red Sandstone basin of Europe, and the wedge to the northwest is commonly referred to as the "Catskill delta," for it is classically exposed in the Catskill Mountains of southern Pennsylvania.

Lower Devonian marine rocks were deposited in the Appalachian miogeosyncline and on the bordering shelf (fig. 6.11). These dominantly carbonate units were soon overlain by clastic units, however, as debris was shed into the same trough from the Acadian uplifts and limestone gave way to marine shale and then to thin-bedded marine sandstone. Marginal marine sediments gave way upward to red siltstone and sandstone and finally to conglomerate. As the trough filled, the Catskill "delta" and related facies spread westward (fig. 6.11). The shoreline was also crowded westward as the detritus was dumped into the geosyncline more rapidly than the area was subsiding. Fine-grained clastic

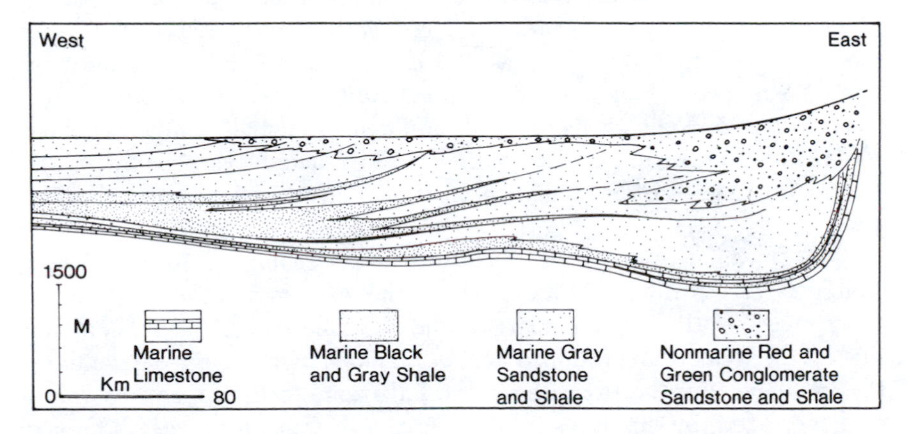

Figure 6.11. Cross section across New York showing Middle and Upper Devonian strata of the Catskill "delta." Facies interpretations of these rocks indicate an eastern source area for the sediments from the Acadian Mountains.

rocks, presumably derived from the orogenic belt, finally spread as far westward as Iowa, reaching there in the latter part of the Upper Devonian.

The clastic wedge reaches maximum thicknesses of over 3 kilometres in eastern Pennsylvania and New York and thins to the south and west. Equivalent beds are only approximately 100 metres thick in central Ohio. The thick clastic section is clearly exposed in the northern part of the ridge and valley province of the Appalachian Mountains, but Devonian rocks are very thin in the southern part of the province. This suggests that no prominent uplifts or source areas were present opposite the southern part of the mountains at that time. There is some suggestion that the southern part of the piedmont was part of Africa and was not welded to the North American continent as a source area until Mississippian time, when the Theic Ocean was closed.

Granitic intrusions are common throughout the Acadian belt. Most of these are postorogenic, although many have radiometric dates of 340-370 million years. The batholiths and stocks appear to be part of a magmatic event that extended into the Mississippian. Some of these intrusions apparently have dates as young as Jurassic.

DEVONIAN TECTONICS

The proto-Atlantic Iapetus Ocean, which had separated northern North America from Europe, was closed from the north southward during the Caledonian and Acadian orogenies. The southern part of the ocean was not closed, however, until the southwestern part of the European plate and the North American plate collided in the Devonian. Continent to continent fusion took place from Greenland and Newfoundland southwestward into New England, and refolded principally eugeosynclinal rocks along the margin of the continent. The Theic Ocean separated the North American-European plate from the African plate during the Devonian.

The Devonian ended a major episode in the geologic history of North America. The Acadian Orogeny along the eastern margin, the Antler Orogeny along the western margin, and the Ellesmerian Orogeny along the northern margin produced a mountainous borderland surrounding the continent. For the first time in the Paleozoic, the continent was more or less ringed by uplifts and these major sediment sources supplied debris that was swept toward the center of the continent, rather than away from it.

7

Mississippian

TOPICS

HISTORY OF STUDIES

The Carboniferous Period, named by Conybeare and Phillips in 1822 for rocks in England, is subdivided into two separate periods in America. These two periods are roughly equal to the Lower and Upper Carboniferous of Europe and Asia, and they are called the Mississippian and Pennsylvanian periods. Henry Schaler Williams proposed, in 1891, that the Carboniferous System in North America be separated into two series, the lower dominated by calcareous rocks and the upper by the coal-bearing clastic rocks in eastern North America. The Mississippian and Pennsylvanian were elevated to systems by Chamberlain and Salisbury in 1906, with the type section of the Mississippian System along the upper Mississippi River valley and the Pennsylvanian from Pennsylvania where the coal-bearing beds are well differentiated. Chamberlain and Salisbury proposed this nomenclature because faunas and floras of the two systems are different and because Pennsylvanian rocks rest unconformably over Mississippian and older rocks across much of the interior of North America, as a result of sea withdrawal from the continents on a worldwide scale (fig. 4.7).

MISSISSIPPIAN STRATA ON THE CRATON

Oldest Mississippian rocks on the eastern craton are shale eroded from the Devonian Acadian uplift on the eastern margin of the continent. These clastic sediments are overlain by widespread Middle Mississippian limestones over much of the interior of the continent. These limestones represent one of the last major carbonate deposits upon the craton and stretch almost uninterrupted from the Cordilleran miogeosyncline of western North America to the Appalachian miogeosyncline (fig. 7.1). Most of the limestone beds are particularly rich in crinoid fragments, particularly the disclike circular columnals of the stem. Crinoids were apparently abundant and widespread animals on the Mississippian sea

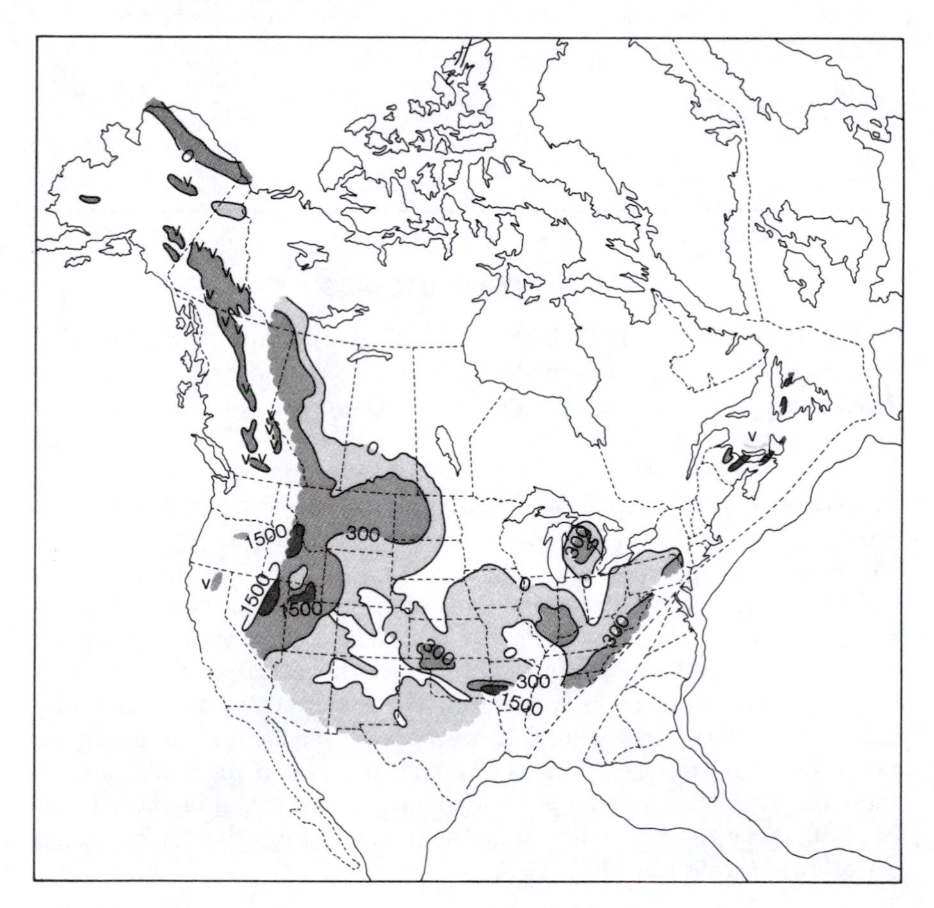

Figure 7.1. Thickness map of Mississippian system in North America. Thickness figures in metres with thickest deposits shown by heaviest shades.

floor, and some secreted calcareous segmented stems several metres long. Fragments of these fossils and others, like brachiopods, corals, and bryozoans, occur in most units. Almost any exposure along a river bank or road cut where Mississippian beds are exposed yields abundant fossils. Mississippian fossiliferous limestones are quarried in Indiana for building stone and are used as a finish stone thoughout the United States.

Upper Mississippian strata throughout the interior of North America consist of interbedded sandstone, shale, and limestone. These rocks record a somewhat rhythmic advance and retreat of the shoreline as the vast expanse of marine water covering the craton withdrew from the craton, perhaps in response to uplifts along the eastern, western, and northern margins of the continent. Shale and sandstone deposited in the upper part of the system were derived from the east and north, as sedimentation in the interior of the continent reflects the influence of Devonian uplifts at the beginning of the period and the minor uplifts that appeared as forerunners of the Pennsylvanian deformation to follow.

The Williston Basin in North Dakota and Saskatchewan and adjacent areas was the site of evaporite accumulation during part of the Mississippian and appears much like the Michigan Basin of earlier times (compare figs. 7.1 and 5.1).

In a general way, Mississippian strata preserve a record of a general regression of marine waters from the craton withdrawing into the marginal geosynclines. Again in a broad view, this completes the cycle from the major transgression of Early Devonian followed by a regression during Middle and Late Mississippian time (fig. 4.7).

CHERT DEPOSITION IN MISSISSIPPIAN ROCKS

Chert is a cryptocrystalline variety of almost pure silica, or SiO_2. The chemistry of silica is unusual, for it precipitates only at very high values of acidity or alkalinity, conditions that would seem to be rare in nature. Silica is carried in solution by most streams in amounts averaging less than 0.0001 percent by weight. The modern ocean into which these streams empty contains even less silica in solution, for most such silica is precipitated by chemicals and biochemical processes.

Much of the dissolved silica is extracted from the sea by sponges or by microscopic organisms, such as radiolarians and diatoms, which use it as the building material of their lacy, glassy skeletons. Most chert seems to be the result of inorganic precipitation. Source for the silica is likely silicic volcanic tephra, or volcanic ash, dust or debris that erupted from volcanic island chains bordering the continent.

Chert is common in many limestone formations of the continental interior, especially in the southeast part where chert beds are up to 60 metres thick. Over most of the interior and within the western miogeosyncline, chert occurs as nodules several tens of centimetres long, or as continuous ribbons of chert a few centimetres thick, regularly interbedded within the limestone though several hundred metres of thickness. In spite of its abundance, chert is among the least understood of sedimentary deposits. Chert illustrates the complexities that exist in natural geochemical systems.

MISSISSIPPIAN STRATA IN THE GEOSYNCLINES

Lower Mississippian rocks accumulated in eastern North America in the Appalachian miogeosyncline and in separate isolated basins, termed epieugeosynclines, for they formed on top of folded earlier eugeosynclinal rocks in southeastern maritime Canada. Lower Mississippian rocks in the Appalachian trough from Pennsylvania southwestward to Virginia are course conglomerate and sandstone and are one to two kilometres thick. These rocks thin westward and become finer grained as well. Nonmarine sandstone in the east grades to marine sandstone and shale in Ohio and finally to limestone in Indiana and Illinois. During the Middle Mississippian the southern part of the trough subsided sufficiently rapidly that marine conditions invaded the trough. The associated limestone and shale transgressed over the distal edges of the former deltas that had developed westward from the Acadian land mass. Renewed uplift in the source area in later Mississippian supplied debris enough to fill the trough above sea level and the shorelines moved westward again. This late uplift may have been produced as Africa and North America moved closer together.

Extensive swamps grew on some of the deltaic margins and the organic debris is preserved as interbedded coal and shale, in limited basins in Upper Mississippian rocks. Coal from Mississippian beds near Roanoke, Virginia, fired the boilers for the *Merrimac* in its fight with the *Monitor*. These eastern coals were exploited long before it was possible to mine the thicker, better coals of the Pennsylvanian in the Appalachian Plateau region to the west.

Epieugeosynclinal basins in maritime Canada were initiated in late Devonian and early Mississippian time. Early Mississippian rocks are principally red conglomerate and sandstone and record the gradual submergence of the several isolated troughs. Rock textures coarsen toward the margins of each of the basins, indicating local sources. Middle and

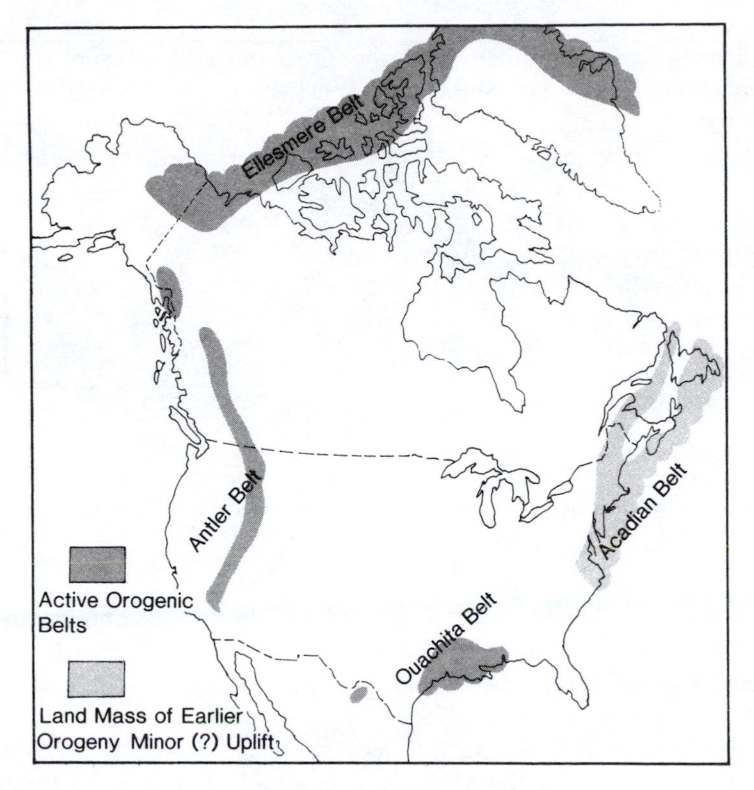

Figure 7.2. Regional map showing orogenic areas in North America during the Mississippian period.

Upper Mississippian beds became marine, as the rate of subsidence exceeded the rate of infilling, so that limestone and some evaporites accumulated at least along the trough of each basin. Basalt and other volcanic rocks are rare, but do suggest that the basins may have opened as weak rifts.

A thick section of Middle and Upper Mississippian clastic debris accumulated in the Ouachita geosyncline (figs. 7.2 and 7.3), and is now exposed in the northern folds and thrust fault blocks of the Ouachita Mountains of Arkansas and Oklahoma. Lower Mississippian rocks of the region consist of only the Arkansas novaculite or chert, and the thin section of these rocks contrasts sharply with the coarse-grained younger beds (fig. 7.3). The latter were derived from the south, possibly from an independent island arc or from uplifts produced by interacting of North and South America.

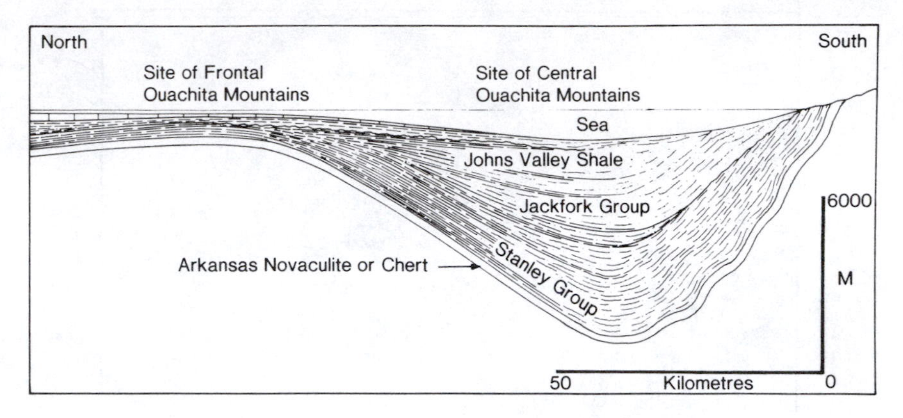

Figure 7.3. Cross section showing Mississippian and Lower Pennsylvanian strata in the south central part of the United States. The source area for these sediments was to the south. Lower Mississippian chert is overlain by Middle and Upper Mississippian clastic sedimentary rocks.

Sediments within the western Cordilleran miogeosyncline also record an adjacent orogenic belt, and that debris was swept eastward onto the cratonal margin. The Antler Orogeny began in the Devonian and continued to be an active source area during Mississippian time, providing great thicknesses of debris to the miogeosyncline (figs. 7.2 and 7.4), as for example in eastern Nevada, where hundreds of metres of sandstone and shale acumulated east of the uplift. Limestone and shale accumulated in the early Oquirrh Basin in Utah, where thick sections were deposited generally east of the Antler clastic apron.

Volcanic rocks, chert, and minor interbedded limestone accumulated in thick sections in the western eugeosynclinal trough, west of the Antler orogenic land mass. The eugeosynclinal rocks are part of volcanic island arcs, and some were deposited in shallow marine water, for corals occur with profusion in some limited beds in western Canada. These complexes accumulated above active subduction zones and, like those of the younger Pennsylvanian and Permian, may be part of separate island arcs that collided with North America in the late Paleozoic or early Mesozoic.

One interesting aspect of Mississippian rocks in the western miogeosyncline is the cliff-forming character of many of the limestone formations. Massive, thick, gray-limestone cliffs characterize Mississippian strata in the Canadian Rocky Mountains, as well as in the Rocky Mountains of Montana, Wyoming, Utah, and Colorado. Part of the same limestone sheet has eroded to produce the prominent Redwall Limestone cliffs on the flanks of the Grand Canyon in Arizona. To the informed traveler

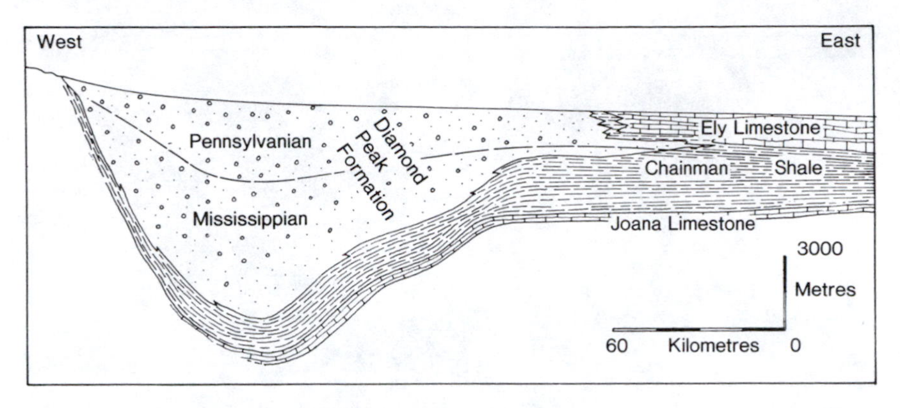

Figure 7.4. Cross section of Mississippian strata in eastern Nevada. Facies patterns indicate the Antler Mountains as a source area during Late Mississippian and Early Pennsylvanian time.

these massive cliffs serve as familiar landmarks in the geologic record in western North America.

The Ellesmerian Orogeny affected the northern part of the Arctic archipelago of Canada (fig. 7.2) and, like the Antler Orogeny to the south, continued from Devonian into Mississippian time. The orogeny involved rocks of both the miogeosynclinal and eugeosynclinal part of the continental border and welded this early sedimentary basin onto the continent. Latest Mississippian and younger rocks are deposited unconformably over the lower Mississippian and older rocks, and are part of the early history of the Sverdrup Basin.

MISSISSIPPIAN LIFE

The Mississippian Period has been called the "Age of Crinoids," referring to a type of echinoderm, like those illustrated on figure 7.5, because of the ubiquitous occurrence of disarticulated crinoid fragments, particularly columnals, found in most Mississippian limestones. Complete crinoids are treasured by collectors and are usually rare in most outcrops, although stem fragments seem everywhere in limestone units.

Foraminifera became important rock-forming fossils in the Mississippian System. Other common fossils in Mississippian limestone are bryozoans, whose lacy fronds cover large areas of rock surfaces. Blastoids, another type of echinoderm, are particularly abundant in the Mississippian. Mississippian brachiopods are somewhat transitional from Devonian

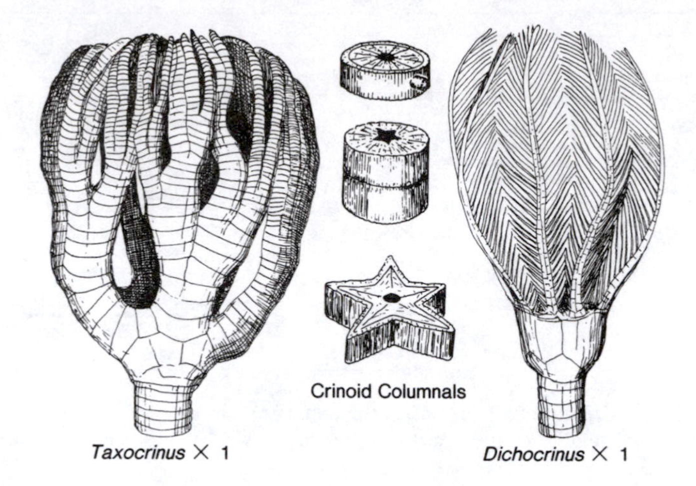

Figure 7.5. Two examples of Mississippian crinoids. Crinoids ("sea lilies") like these flourished in shallow cratonic seas during the Mississippian period. Lifesaver or buttonlike individual stem fragments, like those shown, are the most commonly preserved bits of these fossils.

spiriferids, and to larger spiny productid brachiopods, which are common in the Pennsylvanian and Permian Systems.

Land plants resembled earlier Devonian forms and yet were transitional to the great expansion in plants recorded in Pennsylvanian strata.

MISSISSIPPIAN TECTONICS

The pattern of tectonic activity during the Mississippian Period in North America is almost a direct continuation of patterns established during the Devonian, except for initiation of geosynclinal development of the Ouachita trough along the southern margin and for initiation of epieugeosynclinal basins in southeastern Canada and northeastern United States. Final closure of the Theic Ocean in Late Mississippian and Pennsylvanian resulted from a continuation of the convergent motions of the plates and ultimately resulted in construction of Pangaea, the supercontinent of the late Paleozoic.

The interior of the continent appears to have been remarkably stable, although orogeny affected parts of all the marginal mobile belts, and the western part of Canada, in particular, showed leading edge characteristics.

8

Pennsylvanian

TOPICS

For approximately 45 million years the central part of North America had a remarkably uniform, near sea-level environment. During this interval much of the coal on the continent formed, and the proto-Atlantic Theic Ocean basin closed to suture together the super-continent, Pangaea (fig. 8.1). The margins of the continent, however, underwent significant modification. Sections of the eastern, southern, western, and northern geosynclines were affected by orogeny, and the southwestern and western part of the craton also experienced abrupt local uplifts and deep basins.

Lower Pennsylvanian rocks are restricted to the marginal geosynclines, but by Middle Pennsylvanian, a general transgression of marine waters covered the craton. The water depth upon the platform was probably very shallow because Pennsylvanian cratonic deposits, in general, reflect sedimentary conditions which must have been very near sea level over vast parts of the platform.

CYCLIC DEPOSITION ON THE PLATFORM

In the middle and eastern part of the craton, especially in Illinois, Kansas, and Ohio, Pennsylvanian rocks show rhythmic patterns. A regular sequence of approximately ten thin distinct sedimentary layers that include both marine and nonmarine facies (fig. 8.2) were repeatedly

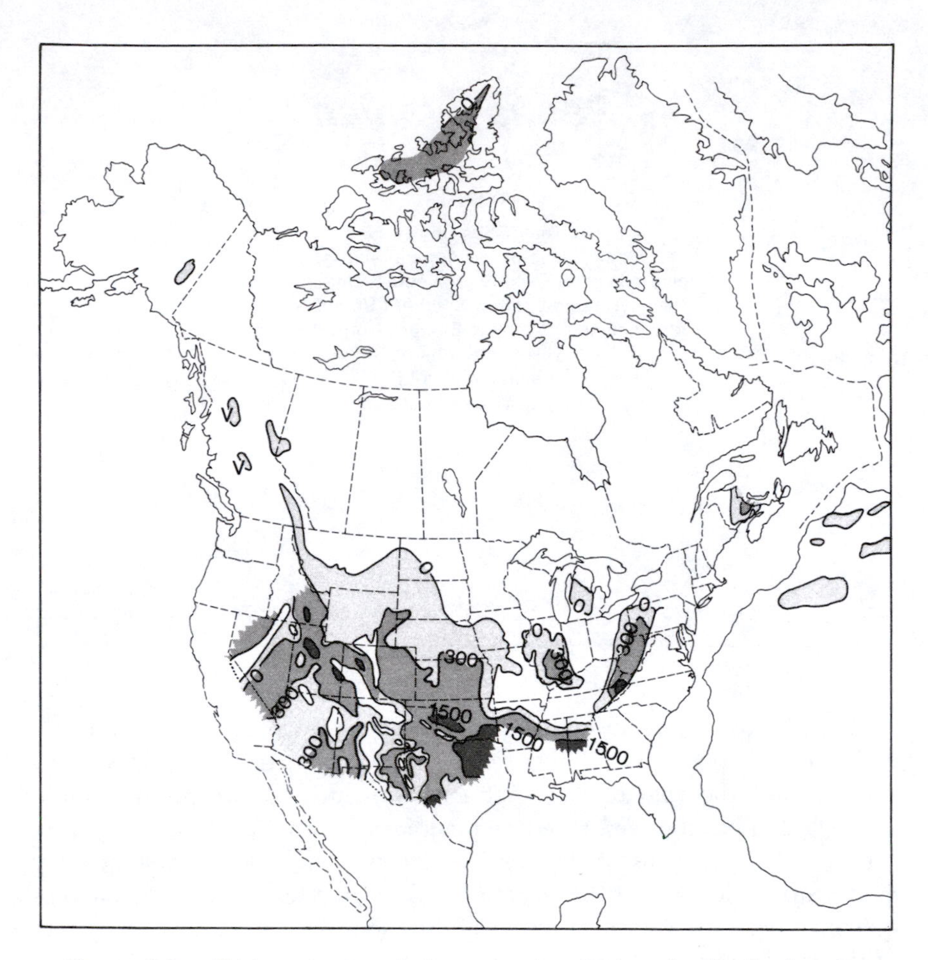

Figure 8.1. Thickness map of Pennsylvanian System in North America. Thickness figures in metres with thickest deposits shown by darkest shades. The irregular patterns reflect the complex orogenic history of the Pennsylvanian period.

deposited. The rock layers consist of continental coal-bearing sandstone and shale overlain by thin, marine, sometimes fossiliferous limestones and shales. The sequence of different rock units may repeat in varying degrees of completeness 50 to 100 times in the relatively thin section. These sedimentary cycles, termed cyclothems, reflect the back-and-forth migration of the shoreline across the near sea-level interior of the continental platform. Cyclothems in Pennsylvania and Ohio show a dominance of nonmarine conglomerate, sandstone, and shale over only thin interbedded marine shale and sandstone. Toward the west, in Illinois and

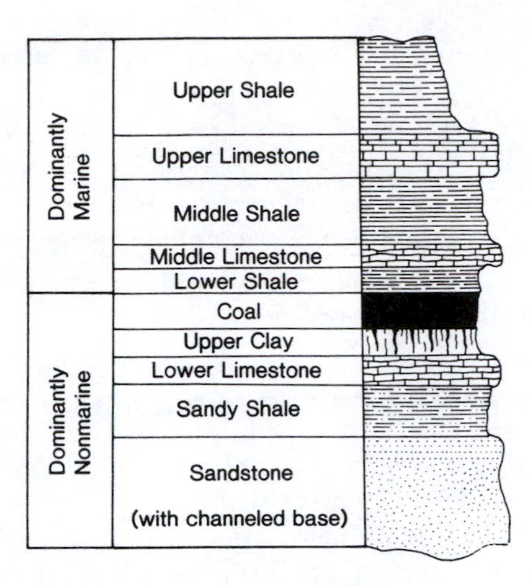

Figure 8.2. Generalized stratigraphic column of a typical midcontinent cyclothem of Pennsylvanian age. This sequence of rock types is cyclically repeated many times in this area, but any individual cycle may not show all rock types. Cyclothems in Ohio and Pennsylvania were dominantly nonmarine, while those in Kansas and areas to the west are more dominantly marine.

Indiana, characteristic cyclothems are like those shown in figure 8.2. Farthern west, in Kansas and Missouri, the cyclothems become more marine, with a generally thin lower nonmarine part to the cycle. Totally marine cycles show in Pennsylvanian rocks in the Oquirrh Basin in Utah.

Cyclothems are more numerous in eastern exposures than they are in western ones, and they do not always correlate great distances across the midcontinent, as was once thought, but many appear to be more local sedimentary packages. During their deposition the shoreline may have migrated back and forth over distances of hundreds of kilometres. This rhythmic shift has been explained recently as a response to erratic movement of continental blocks during plate motion and to associated transgressions of the sea over the continental interior, reflecting the convergence of North America with Africa and Europe as the Theic Ocean closed, followed by regression during periods of tectonic activity. Though on a much smaller scale, this is the same mechanism as the major transgressions driven by orogenic activity at the continental margins, and regression of marine waters during nontectonic periods throughout the Paleozoic. Cyclothems have also been explained as responses of sea-level changes associated with glaciation in the southern hemisphere. As ice

Pennslyvanian plant material represented by the eastern coal fields is enormous.

The rhythmic deposition of cyclotherms has also been explained as a response to lateral shifts of deltaic deposition in the shallow sea. The lower nonmarine part of the sequence represents expanding delta lobes that are buried beneath marine deposits following subsidence of the delta once the stream is diverted into an adjacent low area, in a manner like Cenozoic shifts of the Mississippi delta.

COAL BASINS OF THE APPALACHIAN AREA

Middle and Upper Pennsylvanian rocks in the east-central United States have yielded most of the coal consumed in this country (fig. 8.3). Individual coal beds up to 10 metres thick can be traced for hundreds of kilometres. Coal deposits represent the compressed, partially decayed remains of plant material. A single metre of coal represents the compacted remains of several tens of metres of vegetation. The amount of

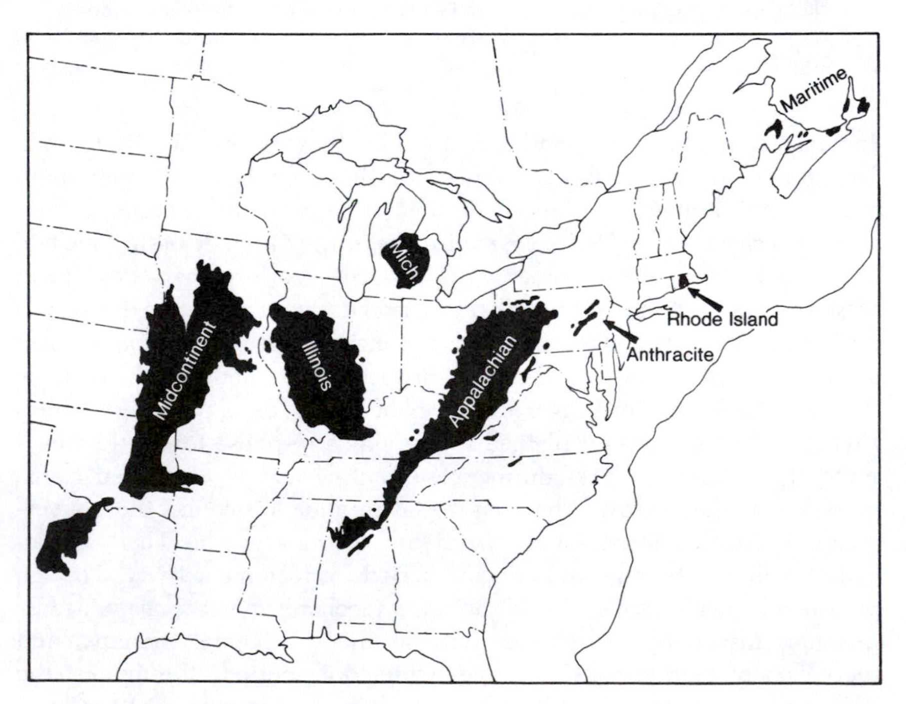

Figures 8.3. Map of eastern North America showing the distribution (black areas) of Pennsylvanian coal fields.

builds up on land, a corresponding drop in sea level reflects a decrease in water in the world's oceans.

The most comparable growths of vegetation in the modern world are found in the coastal swamps along the continents, such as Dismal Swamp in North Carolina, and the Everglades of Florida. A similar environment existed in the vast area presently underlain by coal deposits in eastern North America. Pennsylvanian rocks throughout the world are the prime sources of coal for most of the industrial countries of the world.

UPLIFT IN WESTERN AND SOUTHWESTERN UNITED STATES

The final uplift of the Antler orogenic belt (fig. 8.4) occurred during the Pennsylvanian, ending the activity that started during the Devonian and continued through the Mississippian (fig. 7.4). Coarse sediments were eroded from the Antler Mountains early in the Pennsylvanian, before Upper Pennsylvanian limestone unconformably covered the eroded roots of the orogenic belt.

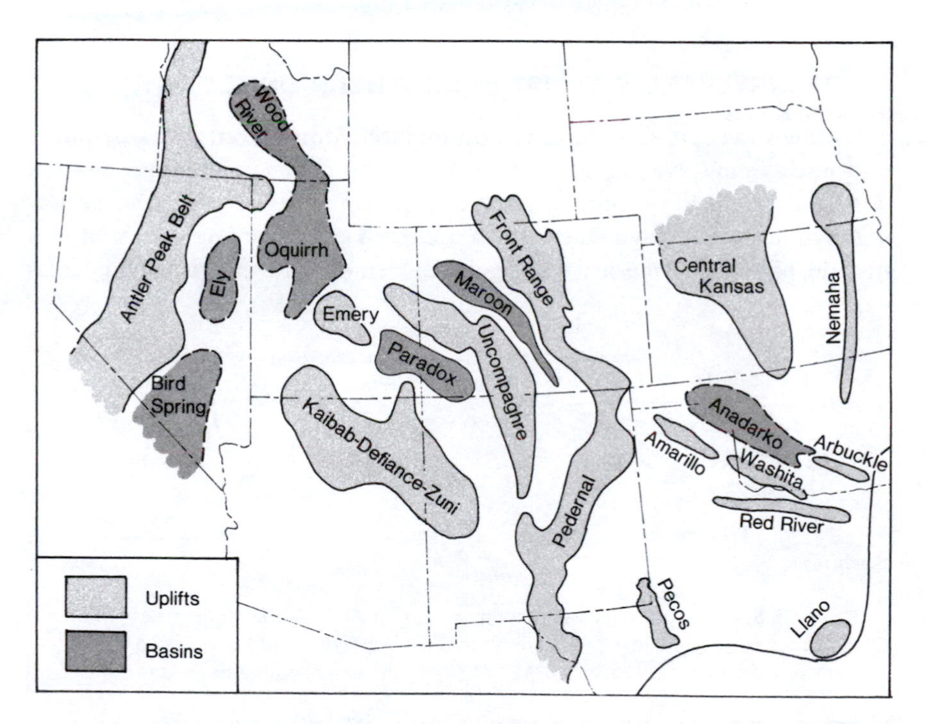

Figure 8.4. Map of Pennsylvanian uplifts and basins of the Colorado Mountains and Antler orogenic belt in southwestern and western United States.

During Middle Pennsylvanian the Colorado Mountains were formed in Utah, Colorado, and New Mexico (fig. 8.4). Although they are sometimes called the Ancestral Rocky Mountains, they are not related to the modern Rocky Mountains except in general location. Deep basins formed between the several uplifts and were usually filled by thick sequences of Pennsylvanian and sometimes Permian clastic strata. Nonmarine conglomerate, sandstone, and red shale accumulated at the margins of the uplifts, although in some basins marine limestone occupied the center of the depressions. The Garden of the Gods in Colorado displays redbeds reflecting the erosion of the Colorado Mountains, as do many of the red cliffs in the Four Corners area and in Canyonland National Park in Utah. Within the Oquirrh Basin of central Utah marine Pennsylvanian and Permian strata total more than 9 kilometres of fossiliferous limestone and sandstone, all in the Oquirrh Formation. These basins and uplifts have a northwesterly trend and are possible indications of an early aborted breakup of Pangaea. The Colorado Mountains continue toward the southeast as the Arbuckle fold belt, where they are terminated by the younger Ouachita Mountains faults and folds.

PENNSYLVANIAN UPLIFT IN SOUTHERN UNITED STATES

A thick wedge of sediments accumulated during Late Mississippian and Early Pennsylvanian in the Ouachita geosyncline, as debris was shed northward onto the continent margin from land masses within a bordering southern eugeosyncline. In the Late Pennsylvanian the southern margin of North America was again deformed and eugeosynclinal ma-

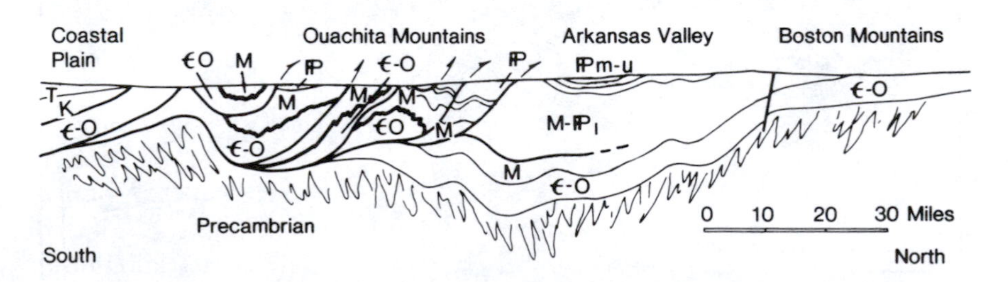

Figure 8.5. Structural cross section across the Ouachita fold belt. Older marginal eugeosynclinal rocks are exposed in the uplifted core of the mountains and younger Paleozoic rocks are involved in folds in the Arkoma basin to the North.

terials were thrust northward onto the miogeosynclinal area. The orogenic pulse formed the Ouachita Mountains in Oklahoma and Arkansas (figs. 7.3 and 8.5), Pennsylvanian and older miogeosynclinal rocks were also folded in the Marathon region of western Texas. The Marathon orogenic belt was at one time continuous with the Ouachita Mountains, for the deformed Paleozoic rocks are mapable in the subsurface around the southern and eastern margin of the Llano uplift in central Texas. By Permian time the Marathon folds were bevelled, resulting in a prominent angular unconformity that separates the folded Pennsylvanian and older strata from the horizontal Permian strata above.

The Ouachita and Marathon deformation marks the initial pulse of a much greater and more widespread tectonic episode in the geologic history of the eastern United States. The orogenic pulse along the southern margin of the continent is interpreted by some as a response to collision of northwestern South America with the North American plate.

ALLEGHENIAN OROGENY IN EASTERN NORTH AMERICA

The Appalachian Mountains, which extend from central Alabama northeastward to central Pennsylvania, were formed in large part during the Alleghenian Orogeny during the Pennsylvanian and Permian. The eastern eugeosyncline had been folded twice already, during the Ordovician Taconic and the Devonian Acadian orogenies. The Alleghenian Orogeny involved principally the miogeosynclinal sedimentary rocks and their eastward transitional margin into older eugeosynclinal rocks. Faulted and folded Precambrian and early Paleozoic rocks form the Blue Ridge and Great Smoky Mountains of the eastern part of the disturbed Appalachian belt. Folded and faulted Cambrian through Pennsylvanian rocks have now eroded to form the Ridge and Valley province of the Appalachian Mountains.

Debris eroded from these newly formed mountains produced a great wedge of Pennsylvanian sediment onto the eastern part of the craton (figs. 8.1 and 8.6). Most Pennsylvanian rocks in eastern North America from Texas to maritime Canada are nonmarine, eroded from the uplifted eastern margin of the continent. These nonmarine beds total 7 kilometres in Arkansas and 4 kilometres in the Canadian maritime region. They are approximately 2 kilometres thick in Alabama and Tennessee. The thick coal beds mentioned earlier are included in these dominantly nonmarine sandstone and shale units, in both the miogeosyncline and on the craton margin.

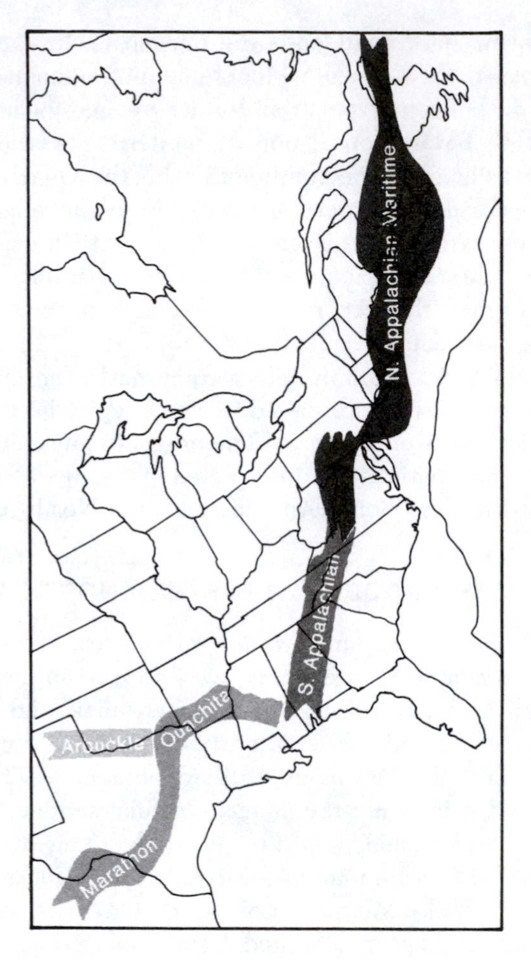

Figure 8.6. Map of eastern and southern North America showing Pennsylvanian and Permian fold belts. The age of deformation decreases northward along the belt.

The Alleghenian Orogeny that affected the Appalachian miogeosyncline represents the last major episode of thrust faulting and folding in eastern North America. The orogeny is generally considered to be a result of collision of the African-South American Gondwana plate with the North American-European plate, closing the Theic Ocean. The collision resulted in the final suturing or welding of that bit of African plate represented by the region east and southeast of the Blue Ridge and Great Smoky Mountains onto the North American continental block.

Just as the Ouachita Mountain trend truncates the Arbuckle folds and faults, so does the southern Appalachian Mountain trend appear to truncate the eastern extension of the Ouachita fold belt. The junction area of the latter, however, is deeply buried beneath younger rocks of the Gulf Coastal Plain. Deformation of the southern part of the Appalachian Mountains appears to have been largely later Pennsylvanian, but the weaker deformation of southeastern Canada appears to have been Permian.

PENNSYLVANIAN ROCKS OF CANADA

Pennsylvanian rocks are recognized only in widely scattered and relatively small areas in Canada and in Alaska. Thick sections of principally nonmarine redbeds occur in the narrowly limited New Brunswick Basin of southeastern Canada. Up to 4 kilometres of sandstone, shale, and coal grade upward into nonmarine beds in the upper part of the section. Pennsylvanian rocks are also very limited in western Canada, where only a few hundred metres of sandstone and siltstone, with interbedded minor carbonate units, are exposed in the Rocky Mountains.

Fossil-bearing Pennsylvanian rocks are even less extensive, although much thicker, in the eugeosynclinal belt along the western margin of the continent. Several kilometres of siliceous sediments, chert, and volcanic debris are known from central and northern British Columbia, in a pattern that persisted from the Devonian to the Permian.

Pennsylvanian stratal successions in Arctic Canada begin with a basal red conglomerate that buries the eroded roots of the Melville orogenic belt. The section soon grades upward into an evaporite, reefoidal limestone, and fine-grained clastic section in the northern part of the basin. Sandstone and conglomerate continued to be deposited in the southern part of the basin into the Permian.

PENNSYLVANIAN LIFE

In the lush coal-forming swamps in eastern America, insects became uncommonly large and abundant. Insect fossils are relatively rare, yet their remains are found in Pennsylvanian strata preserved as delicate imprints usually in soft shale. Spiders, centipedes, scorpions, and winged insects are also found, the largest having a wingspread of up to 75 centimetres.

Marine Pennsylvanian fossils include small protozoans, termed fusulinids, (fig. 8.7), whose test or outer covering was the shape and size

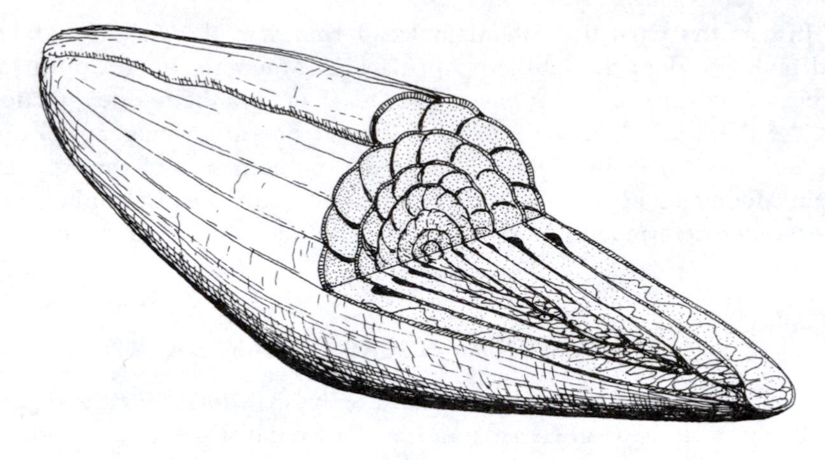

Triticites

Figure 8.7. Cutaway view of typical Pennsylvanian fusulinid, **Triticites,** approximately times 15. These fossils are widespread, abundant, and among the most useful time indicators for Pennsylvanian strata throughout the world.

of a grain of wheat or rice. They became so common that their fossils in some cases represent a major part of the rock. Their pattern and high rate of evolution makes them among the best fossils for dating and correlating Pennsylvanian and Permian strata, particularly in subsurface work relating to oil drilling, where only small samples are recovered.

Brachiopods had evolved into large forms, such as illustrated on figure 8.8. Spiriferid brachiopods, so common in Devonian rocks, are much less abundant in Pennsylvanian strata.

The oldest known fossil reptiles were collected from rocks that filled hollow tree trunks during Lower Pennsylvanian time in Nova Scotia. Our more detailed understanding of the structure of the early reptiles, however, comes from a Permian age fossil called *Seymouria* that represents a structural transition between amphibians and reptiles, for the genus possesses skeletal features otherwise restricted to one group or the other.

Plant evolution during the Pennsylvanian was varied and rapid. Under the extremely favorable conditions that existed in areas of the coal swamps in eastern America, trees grew to heights in excess of 30 metres with trunks up to 2 metres across. The largest trees, called scale trees, are represented by abundant fossil remains, as are the ferns, seed-ferns, and scouring rushes. Leaf and bark impressions of these plants are commonly preserved in association with Pennsylvanian coal-bearing strata

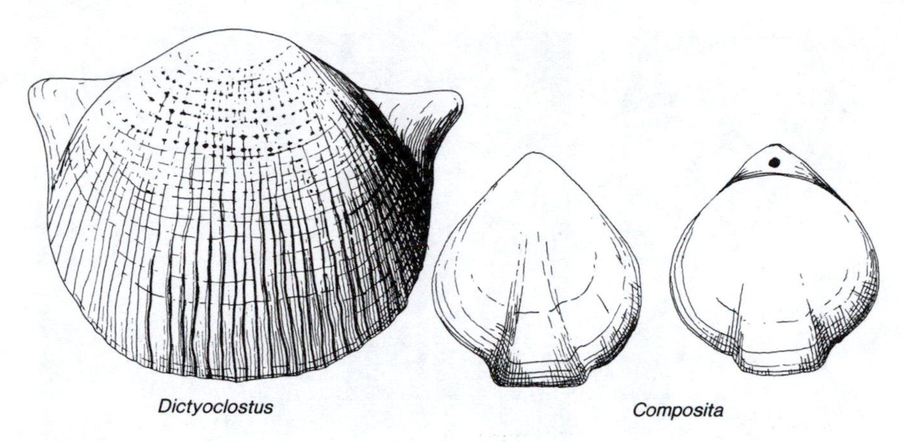

Figure 8.8. Common brachiopods found in Pennsylvanian rocks throughout North America.

(fig. 8.9). Ancestors to the conifers, as well as the first true conifer fossils, are found in rocks of Pennsylvanian age.

PENNSYLVANIAN TECTONICS

Rocks in the Pennsylvanian System record the paradox of one of the most uniform climates known in the Paleozoic from the interior of the continent, yet one of the most structurally active and unstable periods in the Paleozoic of North America. The Alleghenian Orogeny along the eastern margin of the continent reflects the final closure of the Theic Ocean basin and construction of the supercontinent, Pangaea. Folding, faulting, and uplift of eastern and southern North America corresponds in time to the Late Pennsylvanian Hercynian Orogeny in western Europe and Asia. Areas affected by the Hercynian Orogeny include Wales, France, Germany, Bohemia, U.S.S.R., and central Asia. The converging continents behaved, in a general way, as did the converging Asian and Indian continents when they formed the Himalayas during the middle of the Tertiary Period, although in the formation of the Himalayas, the Indian plate was thrust under the Asian plate, rather than simple head-on collision and fusion of the continents, as we see in eastern North America.

The southern margin of North America also experienced compressive stress. Plate relationships are somewhat obscure because the critical rocks are deeply buried beneath the Gulf Coastal Plain. There is some

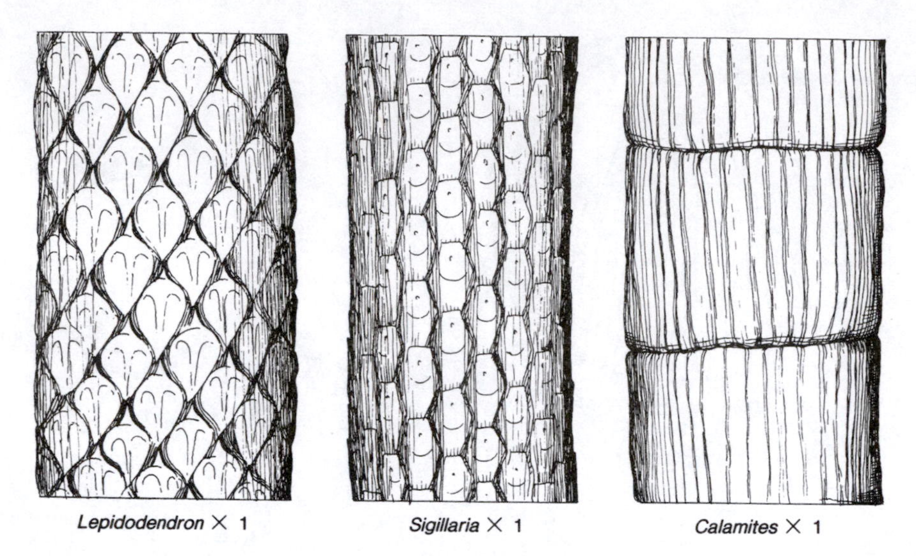

Figure 8.9. Stem sections of typical Pennsylvanian plant fossils showing scars of scales (**Lepidodendron** and **Sigillaria**) and typical segmented growth patterns of **Calamites,** an ancestor of the modern scouring rush. These fossils are particularly abundant in Pennsylvanian coal deposits.

suggestion, however, of collision of South America, or of an independent island arc, into the southern margin of the continent.

Northwesterly trending uplifts and basins within the craton in southwestern United States, as well as a continuation of the epieugeosynclinal basins in maritime Canada, are probably a response to adjustment to horizontal stresses generated by final covergence of the major continental plates into Pangaea or to incipient but aborted rifting. A consistent history of volcanic accumulation indicates continued leading edge subduction along the western edge of the continent during Pennsylvanian time. Volcanic rocks are particularly evident in western Canada.

The end of the Pennsylvanian Period marks a major shift in the geologic history of North America. In the remaining history, recorded in younger rocks, the emphasis of activity shifted to the western part of the continent as Pangaea separated and North American begain moving relatively westward. Major tectonic events were associated with the leading edge of the continent, or west coast, while the trailing edge, or east coast accumulated erosional debris upon the intensely deformed margin. The southern Gulf Coast margin and the northern Arctic Canada margin also took on trailing edge characteristics for much of the remainder of geologic history.

9

Permian

TOPICS

The Permian System was proposed by Sir Roderick Impy Murchison in 1841 for a distinctive series of fossiliferous rocks exposed near the small town of Perm, and elsewhere along the western flank of the Ural Mountains in Russia. Murchison noted that these rocks contain a flora intermediate in appearance between the Carboniferous and Triassic fossil plants of western Europe.

Permian rocks were first documented in North America in 1858 when Major Frederick Hawn, in Kansas, and Benjamin Franklin Shumard, in Texas, recognized and described Permian fossils. Since that time the abundantly fossiliferous reef-related sections of the Glass Mountains and Guadalupe Mountains in Texas and New Mexico have become the standard reference section for North America.

Throughout the world Permian rocks record great environmental extremes when compared to the relatively uniform conditions of the preceding Mississippian and Pennsylvanian. Permian seaways covered less of the continents; glaciation was widespread in the Southern Hemisphere; while in the Northern Hemisphere deserts became extensive, as indicated by widespread sand dunes and salt deposits, and organic reefs flourished in nearby seas. As a result of these changing environmental conditions, many plants and animals became extinct; others adapted and developed toward modern types.

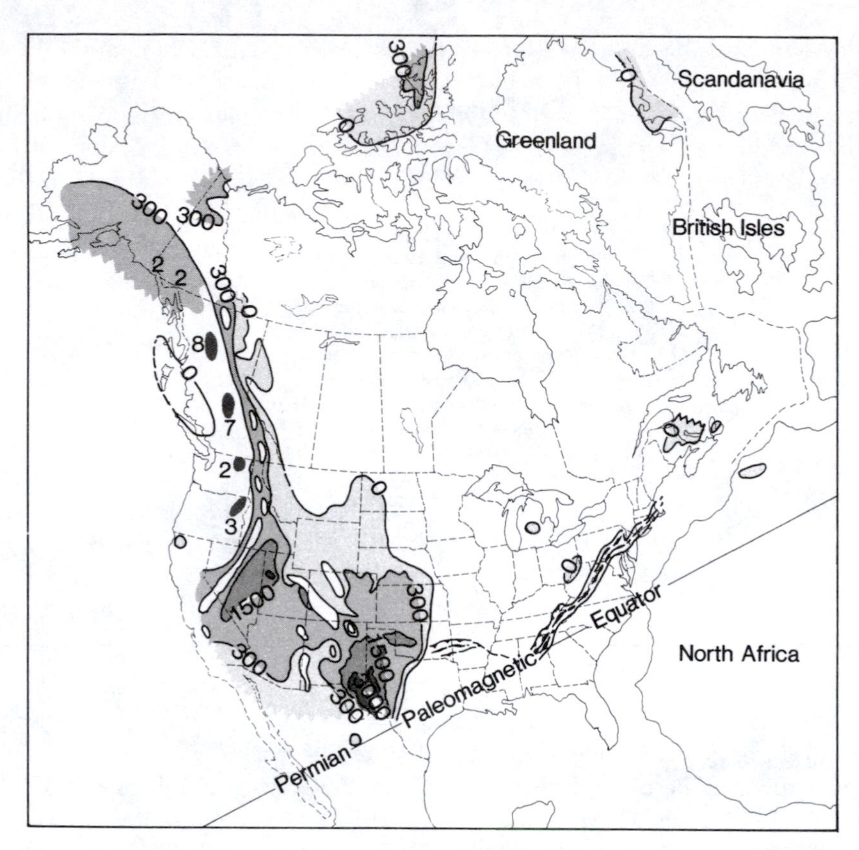

Figure 9.1. Thickness map of Permian System in North America. Thickness figures on lines are in metres; isolated values in western eugeosynclinal belt are in kilometres. Fold and thrust-fault trends in Ouachita area of Oklahoma and Appalachian Mountains are shown schematically.

PERMIAN PALEOGEOGRAPHY

The restricted extent of Permian strata east of the Mississippi River is clearly shown on the thickness and distribution map (fig. 9.1). The entire eastern part of the present continent was highland, either on the eroding roots of the Acadian land mass or formed by the Alleghenian or Appalachian Orogeny of Pennsylvanian and early Permian time. As a consequence, marine Permian rocks are absent east of the Mississippi River. Nonmarine sandy redbeds are preserved, however, in New Brunswick and Prince Edward Island in the New Brunswick Basin, and to the southwest in the center of the Michigan Basin and along the trough of the Dunkard Basin in western Pennsylvania, eastern Ohio, and West Virginia. In the west, Permian rocks are widespread and record that Early Permian marine conditions gave way to nonmarine deposition in the Middle and Upper Permian from northern Mexico and Texas northward to North Dakota. Only along the present western margin of the continent did marine conditions prevail until near the end of the period in a belt from northern California, through western Canada, to Alaska. Lower Permian beds in the Canadian Arctic Archipelago spread across beds folded during the Melville Orogeny; during the later Permian marine deposits accumulated in the Sverdrup Basin on that northern part of the continental margin.

ALLEGHENY AND NEW BRUNSWICK BASINS

In the Allegheny basin, rocks of Early Permian age are similar to those of the Late Pennsylvanian; both represent continental deposition in stream, swamp, lake, and delta environments near sea level. Permian rocks reach a thickness of 300 metres in the Allegheny area where they are called the Dunkard Group. Dunkard rocks include cyclically bedded mudstone and standstone, with thin interbedded conglomerate freshwater limestone and coal. Coarsest rocks are on the southeast, suggesting that the source was a land area in Virginia and adjacent states. Fossils from Dunkard strata are chiefly plants of Early Permian age. Middle and Late Permian rocks are absent from the area.

Early Permian deposits in the New Brunswick Basin of maritime Canada, as in the Allegheny Basin, continue to reflect environmental patterns established in the Pennsylvanian. The upper part of the Pictou Group, dated as Early Permian on the basis of plant spores, consists of up to 2.4 kilometres of gently folded red sandstone, siltstone, and shale, with minor gray sandstone, mudstone, and some coal. It represents continental deposition of material derived from low adjacent uplifts.

PERMIAN FOLDING IN THE APPALACHIAN MOUNTAINS

We can place a definite upper limit on the age of the Alleghenian folding because Upper Triassic beds were deposited unconformably across erosionally bevelled Appalachian folded rocks. The lower limit for the folding is more difficult to date precisely. Lower Permian rocks in the Allegheny basin are conformably folded with Pennsylvanian strata; therefore, at least part of the folding must be of post-Pennsylvanian age. However, these folds lie west of the main Appalachian folds. In the main

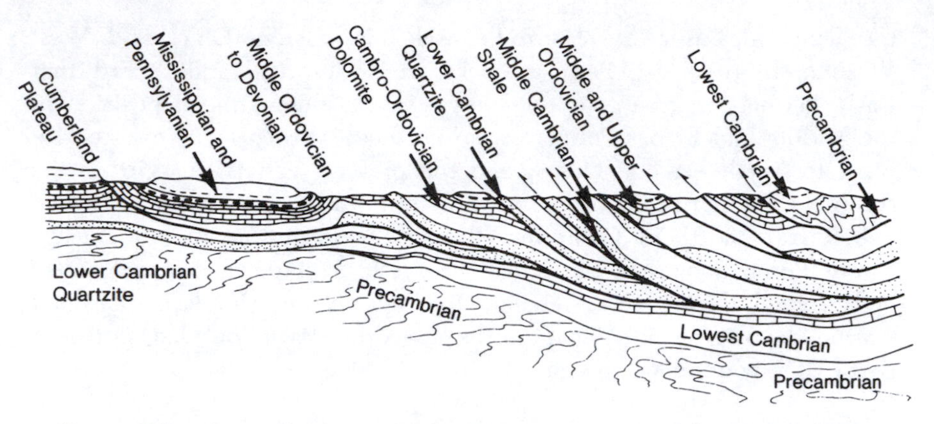

Figure 9.2. Generalized structural cross section across the Appalachian Mountains showing the thrust faults of the southern part of the Alleghenian orogenic belt. The ridge and valley province is composed of imbricate slices of Cambrian and younger Paleozoic formations. Late Precambrian and Cambrian rocks are exposed in the Smoky Mountains to the east.

fold belt, Pennsylvanian strata are the youngest rocks actually involved in folding. Folded Mississippian and Pennsylvanian strata in Massachusetts and Rhode Island extend the Alleghenian fold trend toward maritime Canada. Early Permian strata in southeastern Canada are only gently folded and lie across sharper folds of earlier strata, suggesting that northwest compressional forces diminished in strength by Early Permian time.

Alleghenian folding developed by episodes during Late Mississippian, Pennsylvanian, and Early Permian; slow growth of these folds is suggested by the observation that Late Paleozoic strata are thicker in downfolded areas and thinner in upfolds as a result of erosion and deposition acting at the same time as folding. The orogeny ceased well before Late Triassic time because nonfolded Upper Triassic strata rest with angular discondance upon folded Paleozoic strata. Thrust sheets are the dominant structures in the southern part of the Appalachian Mountains; folds dominate in the northern Applachians in the United States. The Alleghenian Orogeny has been variously interpreted as produced by "pistonlike" compressive stresses resulting from continental collision or as almost slumplike sheets, which moved down off a major upwarp that heralded development of a new spreading center and breakup of the supercontinent Pangaea.

Postorogenic granite intrusions cut strongly folded and metamorphosed Carboniferous rocks in the isolated epieugeosynclinal basins and

Figure 9.3. Restoration of approximate relationships of continental elements within the supercontinent Pangaea during Permian time.

older rocks of Rhode Island, Massachusetts, New Hampshire, and Maine. These intrusions have been dated as ranging from 260-180 million years old, indicating emplacement during the Permian and possibly as recently as Jurassic.

PANGAEA-PERMIAN SUPERCONTINENT

If we work backward from the present position of continents, using motions derived from sea-floor spreading and paleomagnetism, we can reconstruct a series of earth models with some success back to Permian time. Figure 9.3 illustrates the Permian positions of North America, western Europe, and Africa. The Permian geology of North America and North Africa lends itself to reconstruction, since there are similarities in the Permian record in the two continents. As in eastern North America, there is little Permian rock in North Africa, substantially less than Pennsylvanian deposits. Thin continental Permian redbeds lie above bevelled folded Mississippian and Pennsylvanian strata, and are in turn overlain by unfolded Triassic rocks, a general relationship we have already noted in maritime Canada. Trends in the Pennsylvanian Hercynian folds in North Africa are north to northeast, essentially parallel to Alleghenian or Appalachian fold trends in North America.

NORTH AMERICAN CONTINENTAL INTERIOR

Figure 9.1 shows extensive Permian deposits in states west of the Mississippi River. Although in much of this area the Permian is characteristically composed of continental redbeds, at least two major marine inundations extended from Mexico and Texas northward into Canada during Early and Middle Permian time. When the seas later retreated southward, Permian evaporites were deposited in coastal lagoons and basins and formed commercial salt deposits in Kansas. Basins in New Mexico and Texas were sites of thick evaporite accumulation by Late Permian.

PERMIAN REEF-RIMMED BASINS

Permian rocks in West Texas and New Mexico produce petroleum from thousands of wells that have penetrated a reef-rimmed basin complex. Permian deposits are nearly flat above older folded structures formed during the Pennsylvanian in the Texas-Oklahoma area. By Early Permian time, the features were established that characterized the Permian Basin complex in Texas and New Mexico (fig. 9.4). Most striking is the Delaware Basin, which contains Permian deposits up to 6 kilometres thick.

Early Permian sediments on the shelf and platform areas that surround the Delaware Basin on all but the southwest side are light-colored granular limestones and dolomites, contrasting with the black limestones and shales within the basin. We interpret this to indicate that water depth was much greater in the basin, perhaps 300 to 600 metres deeper, and that sediments along the shelf margin were deposited nearly at sea level. In a deep basin such as this, normal circulation and growth of organisms may occur in the surface water, while the water near the bottom is stagnant, permitting the accumulation of organic detritus. The oxygen-poor water at the bottom of the basin cannot support the bacteria

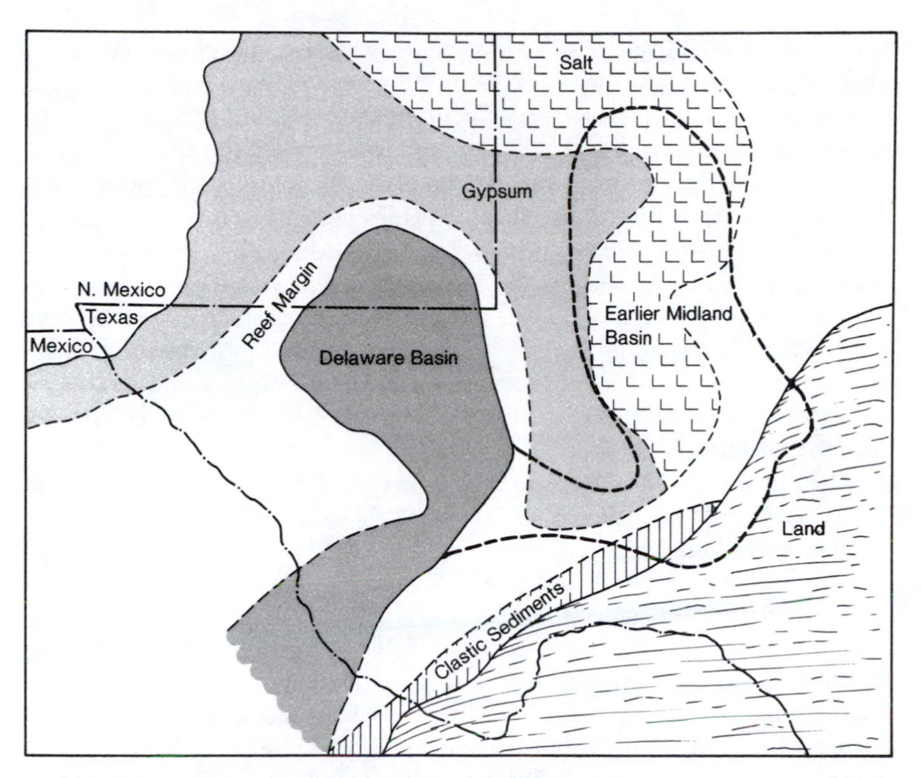

Figure 9.4. Map of Permian basins in Texas and New Mexico. Permian sediments total 6 kilometres in greatest thickness in the Delaware Basin. Dotted heavy line outlines earlier Midland Basin.

that ordinarily consume organic matter, which settles to the bottom from animals living in the oxygenated water near the surface. Under the same circumstances hydrogen sulfide usually develops and produces disseminated black iron sulfide. The iron sulfide and organic materials color the basin rocks dark gray to black.

By Middle Permian time, small patch reefs began to grow on the margins of the basin; these developed into thick unbedded limestone built mostly by calcareous algae and sponges, but also including bryozoans, echinoderms and brachiopods. These massive limestones have been interpreted as reefs for many years, but recent studies suggest that these organism-produced limestones may have accumulated in water a few tens of feet deep, and may not be comparable to modern reefs. As the basin and shelf margin areas subsided, sponges, algae, and other organisms that populated the transition zone grew upward, keeping approximate pace with subsidence. Blocks of the massive marginal limestones broke off and slid into the basin, helping to form extensive talus deposits.

Because sandstones and shales continued to fill the subsiding basin, we conclude that sand and silt were carried by currents from the land areas to the north, across the shelf, and through the areas of sponge and algal growth, into the basin. Shallow-water lagoonal environments persisted in the platform areas behind the basin edge. Here limestone and dolomite were deposited near the basin edge, while farther to the north evaporites were precipitated along the coastline. Carlsbad Caverns in New Mexico have been carved by later ground water solution in these Permian lagoonal and marginal massive limestones. The Guadalupe Mountains of West Texas expose much of these basin to platform relationships, as shown on figure 9.5.

Late in the Middle Permian, water in the Midland basin, another of the West Texas basins shown on figure 9.4, became so shallow that evaporites became the dominant deposits. In Late Permian time the same thing happened in the Delaware basin, and the basin margin organisms were killed by the changing conditions. Basin subsidence continued at a rate equal to deposition of evaporate salts. Gypsum, salt, and potash deposits of latest Permian age accumulated both in the Delaware Basin area and on the platform areas to the north along the New Mexico-Texas border. Reconstructions of ancient paleogeography show that the paleomagnetic equator crossed northeastward through southern Texas (fig. 9.1), so it is not surprising that evaporation rates were high in these shallow equatorial seas.

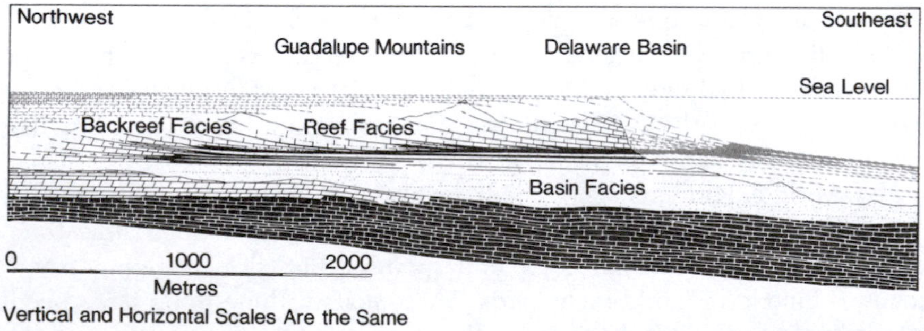

Figure 9.5. Permian lithologic relationships shown in the fault scrap on the western face of the Guadalupe Mountains, at the west edge of the Delaware Basin, in Texas and New Mexico. The "reef" grew upward and basinward over debris as the region subsided. The Delaware Basin, at the right, was filled with evaporites in Late Permian time.

COLORADO MOUNTAINS

Only a few of the Colorado Mountain uplifts that were prominent in the Pennsylvanian in the midcontinent and Rocky Mountain area remained unblanketed by Permian deposits (compare figs. 8.4, 9.1, and 9.6). The largest positive area, the Uncompahgre uplift in Utah, Colorado, and New Mexico, was flanked on the southwest by the Paradox Basin, into which some 2 kilometres of clastic rocks were shed from the uplift, mostly in early Permian time. Other basins and uplifts in nearby areas of the Pennsylvanian Colorado Mountains became subdued, as erosion lowered the uplifts and transported debris filled the basins.

KAIBAB-PHOSPHORIA SEA OF IDAHO-ARIZONA

The simple patterns of the early Paleozoic miogeosynclinal belt in western Utah and Nevada were modified in late Paleozoic time by local basins (as shown on fig. 9.6). The most prominent of these is the Oquirrh

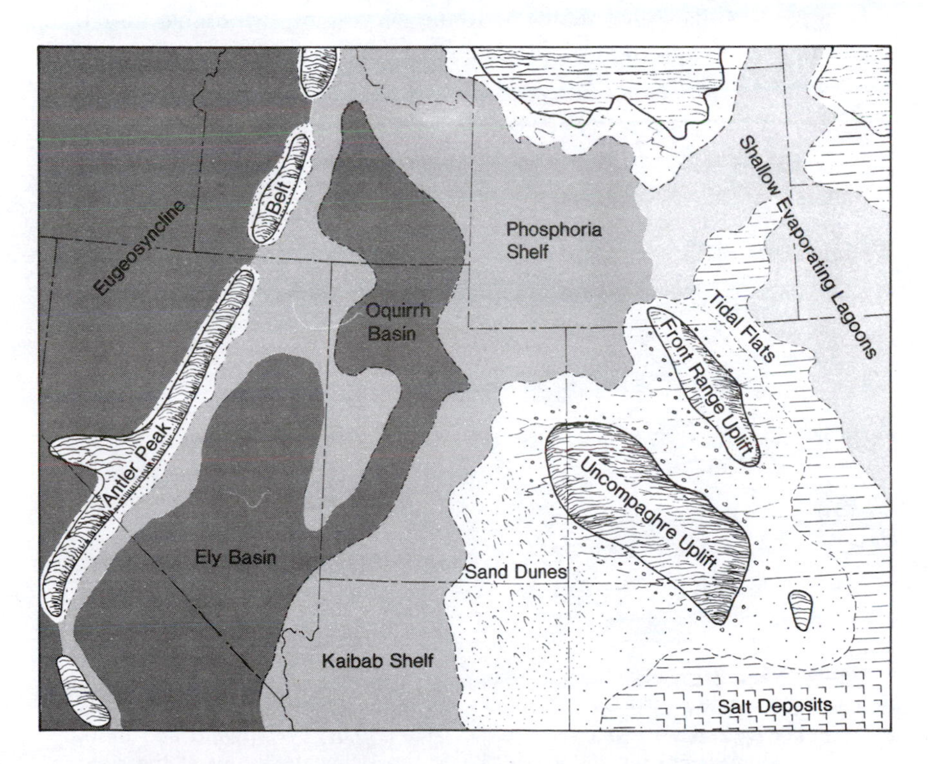

Figure 9.6. Permian basins and uplifts in western United States.

Basin in Utah, in which 4 kilometres of limestone and fine quartz sandstone accumulated during early Permian time. Marine inundation had its greatest extent in the middle Permian and deposited a widespread carbonate sheet across the miogeosyncline and onto adjacent shelves. The Middle Permian Kaibab Limestone now forms the rimrock of the Grand Canyon and is widely exposed over the surrounding plateau area. Beneath it the earlier Permian Hermit Shale, Supai Formation and Coconino Sandstone record continental red bed and sand dune deposition. A cross section of these relationships is shown on figure 9.7. To the east, in the Four Corners area, remnants of Permian DeChelly Sandstone now form the magnificent towers of Monument Valley in southern Arizona. These rocks grade northeastward into totally nonmarine sandy and conglomeratic redbeds in southwestern Colorado and eastern Utah.

At the northern end of the miogeosynclinal area in Utah, Idaho, and Wyoming unusual phosphatic sediments were deposited as the Phosphoria Formation, where upwelling deep water from the west spread out onto the shelf area to the east (fig. 9.6). The Phosphoria Formation is mined for fertilizer phosphates in Idaho, Wyoming, Montana, and Utah. The maximum expansion of the Kaibab-Phosphoria Sea, as indicated on

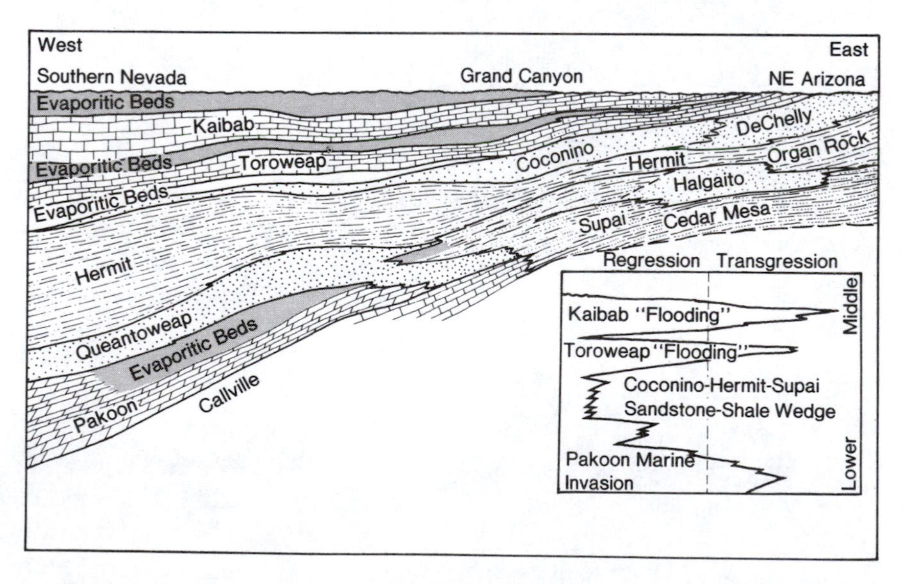

Figure 9.7. Cross section from miogeosyncline (west) to craton (east) of Permian rocks in northern Arizona and southwestern Colorado. Inset shows marine transgressions and regressions schematically. Triassic rocks rest unconformably across these formations.

figure 9.6 corresponds with the maximum development of Middle Permian limestone in many other parts of the world.

WESTERN EUGEOSYNCLINES

The Permian System records a major episode of volcanic activity in western North America. Although later geologic events have broken up, intruded, or concealed the continuity of Permian rocks to the extent that we cannot restore Permian land masses and rock thickness trends in the eugeosyncline with the same certainty that we can in the midcontinent area, nonetheless, certain generalizations are apparent. The Permian eugeosynclinal belt, followed by great thicknesses of unconformably are composed of poorly fossiliferous, predominantly volcanic material. In general they consist of a lower and upper volcanic section separated by a Middle Permian fossiliferous limestone. The Sonoma orogenic belt, along the Antler orogenic trend in central Nevada separates the miogeosyncline from the volcanic eugeosynclinal rocks to the west; this orogenic belt may have extended northward through Idaho and British Columbia.

Three faunal provinces are recognizable in Permian rocks of western Canada. Miogeosynclinal siliceous and dolomitic rocks are preserved in an eastern belt that is exposed in the Rocky Mountains. Two additional faunal provinces that have fossils with Asian affinities are recognizable in the eugeosynclinal belt along the western margin of the continent (fig. 9.8). The western fusulinid coral-brachiopod faunas in northern California, western British Columbia, and southeastern Alaska differ from the faunas found in eastern British Columbia, Idaho, and eastern Oregon, as shown on figure 9.6. The suggestion has been made that rocks of these two faunal provinces represent two originally widely separated island-arc sequences that have been telescoped geographically by plate convergence. Ultrabasic intrusive activity in Late Permian or Early Triassic time in British Columbia, and the general lack of Lower Triassic rocks in the eugeosynclinal belt, followed by great thicknesses of unconformably overlying, andesitic volcanic rocks are evidences of a Late Permian to Early Triassic orogenic disturbance.

ARCTIC CANADA

In northernmost Arctic Canada, a spoon-shaped mass of sediments as much as 12 kilometres thick accumulated in late Paleozoic, Mesozoic, and Cenozoic time to form the Sverdrup Basin. Pennsylvanian clastic rocks in the center of the basin grade upwards into Permian limestones

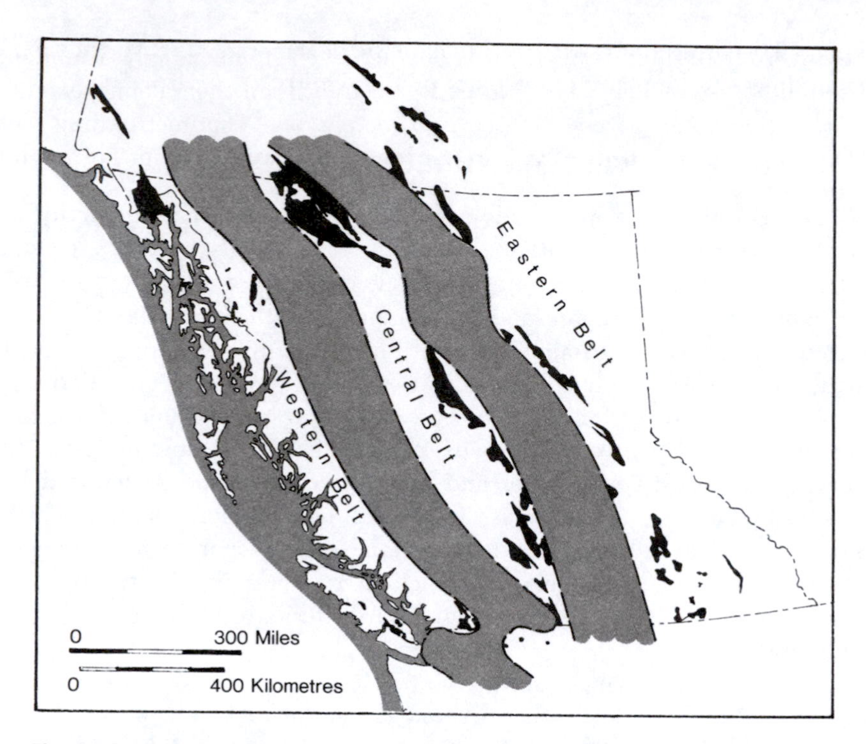

Figure 9.8. Permian faunal provinces in western Canada showing possible separate tectonic blocks in western British Columbia, as based upon differences in faunal composition. The eastern belt is in the miogeosyncline and shelf. The central belt and western belt are both within the eugeosyncline and contain thick sections of volcanic rocks. (After Monger, J.W.H. and Ross, C. A., Canadian Jour. Earth Sci., v. 8, pp. 259-278, 1971)

more than one kilometre thick, or are overlain by Permian clastic rocks and limestones along the basin margin. The Sverdrup Basin trends into the modern Arctic Ocean, which may have been smaller or even nonexistent at that time, according to some reconstructions of the Permian supercontinent Pangaea.

PERMIAN LIFE

For many lineages of organisms, latest Permian time was a crisis. Many groups, such as trilobites, fusulinids, ancient corals, and some types of brachiopods, which had been successful, even dominant, during the Paleozoic, failed to adapt to some environmental change that was widespread near the close of the period, and these groups became extinct.

Invertebrate life in the sea was characterized by amazingly abundant fusulinid foraminifera. These forms were larger than their Pennsylvanian predecessors; most were only approximately one centimetre long but some grew 3 or 4 centimetres long. Some Permian rocks in Texas and New Mexico are formed nearly completely of fusulinid fossils. So useful are these organisms as zone fossils that stratigraphers recognize the base of the Permian as the first occurrence of the fusulinids *Schwagerina* or *Pseudoschwagerina*. In spite of their abundance in Permian seas, fusulinids were also victims of the Permian extinction.

Spiny productid and oysterlike brachiopods flourished during the Permian (fig. 9.9), yet disappeared from the seas near the end of the period. Typical Paleozoic corals also became extinct, as did the trilobites, which had been steadily declining since their heyday in Cambrian and Ordovician seas.

From their origins in the Pennsylvanian Period, reptiles increased in numbers and kinds during the Permian, to begin their dominance among vertebrates, which lasted through the Mesozoic Era. Great fin-backed reptiles approximately 1 metre high (fig. 9.10) characterize Permian forms. One group of Permian reptiles, the theriodonts, evolved complex differentiated teeth, and became the forerunners of mammals, which first

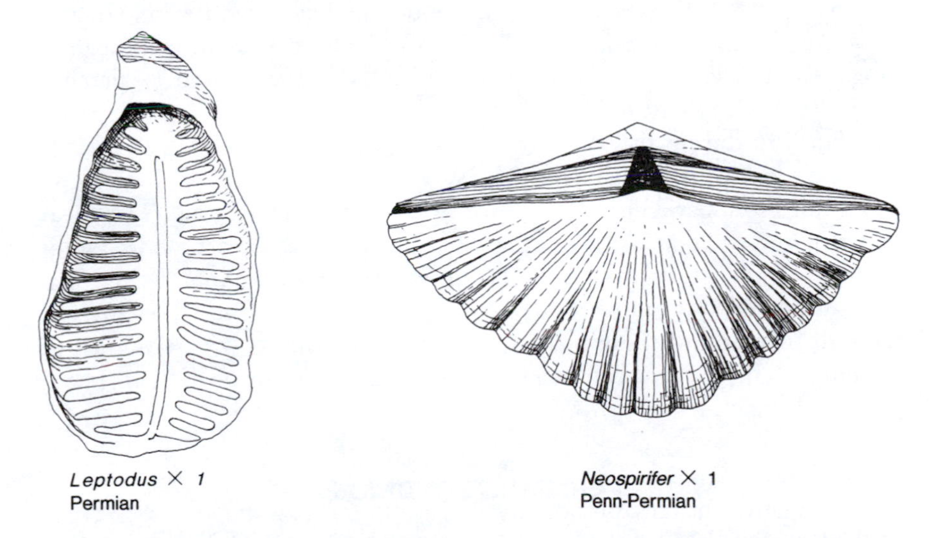

Figure 9.9. Permian brachiopods. **Leptodus** is an unusual brachiopod having one shell with slits. It is found mainly in reeflike mounds around the Texas basins. **Neospirifer** is a more ordinary type of brachiopod and is widely distributed in Permian marine strata.

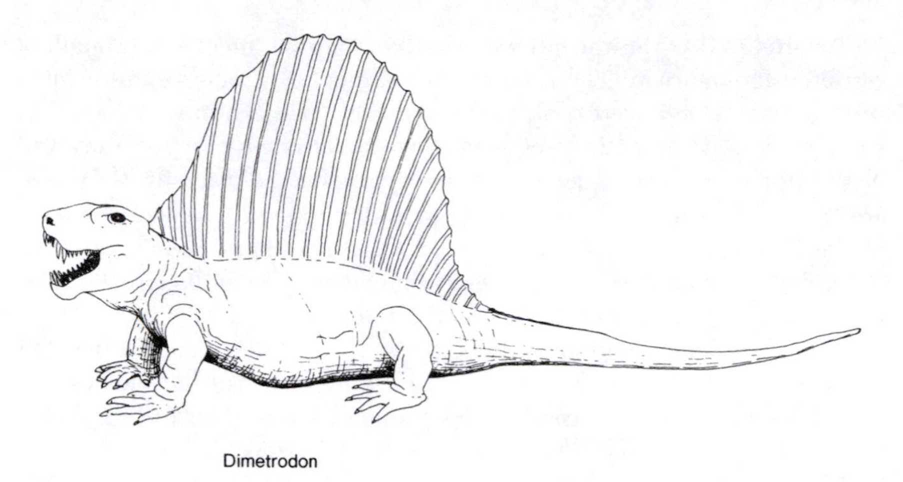

Figure 9.10. Permian "fin-backed" reptile, **Dimetrodon,** was about 2 metres long with the "fin" rising about 1 metre above the backbone. **Dimetrodon** was the largest land carnivore of Permian time. The function of the fin is a puzzle, but is thought to have been a thermoregulating device. The thin "sail" would have allowed rapid heat dissipation after vigorous activity.

appear in the Triassic Period. Theriodonts, whose fossil remains are best known from South Africa, represent a link between reptiles and mammals. Among the vertebrates the Permian was also a time of mass extinction. Seventy-five percent of Permian amphibian families and 80 percent of the reptile families failed to survive the end of the era.

Plant life during the Permian shows a decline in numbers and kinds when compared to the diverse Pennsylvanian types. This is very likely the result of climatic changes toward aridity in Permian time. The great scale trees, *Lepidodendron* and *Sigillaria,* nearly became extinct. A group of seed ferns, *Glossopteris* and *Gangamopteris,* are found in Permian strata from every continent in the Southern Hemisphere. The present wide distribution of these plant fossils is one of the lines of evidence of breakup and spread of continental fragments as visualized by early proponents of continental drifting.

PERMIAN TECTONICS

During the Permian, the eastern and western parts of North America reflect two very different plate relationships. Widespread Permian volcanic rocks in the western eugeosyncline suggest that island-arc systems like those present throughout earlier Palezoic periods continued above active subduction zones along the west coast.

Folding in the Appalachian area and a lack of Permian volcanic rocks along the east coast suggest that we are dealing with interaction between two continental plates, probably Africa and North America, as part of Pangaea. General folding of the Appalachian Mountain belt is evidence of the compressive stress to which the east coast of North America was subjected as Pangaea was formed, or as gravity type glide folds produced from a major upwarp. Development of small local basins and uplifts during the late Paleozoic in maritime Canada may reflect differential movement between minor blocks caught along transform faults between North America and Europe, like minor blocks along the modern San Andreas fault system in California. They may also reflect tensional stress related to partial disruption of the North American-European block prior to creation of Pangaea and closing of the Theic Ocean.

10

Triassic

TOPICS

Break-Up of Pangaea
Western Interior
Alberta-Northeastern British Columbia
West Coast Eugeosyncline
Alaska-Yukon
Arctic Islands
Triassic Life
Triassic Tectonics

Triassic geologic history is a record of the final days of Pangaea and its initial break up. From latest Triassic through the balance of the Mesozoic and Cenozoic Eras, the supercontinent has continued to fragment. When Pangaea started to split in Late Triassic time, it had been in existence nearly 150 million years. As North America and Africa began to separate, the rift that opened between them was the ancestral North Atlantic Ocean.

The North American Pacific Coast was the trailing edge of the continent as Pangaea formed and had been the site of marine sedimentation more or less continuously since the Precambrian. As the continent began its westward movement in the Mesozoic, the west coast became the leading edge and was the site of both Mesozoic and Cenozoic tectonism on a grand scale.

It was in Triassic strata that the early founding geologists recognized a profound change in the kinds of fossils from those preserved in Permian rocks. Half of the fossil families known in Permian rocks are not found in the Triassic. This fact formed the basis of separating the Paleozoic and Mesozoic Eras at the Permian-Triassic boundary. Figure 10.1 illustrates the correlation of the construction, existence, and break up of Pangaea, and the patterns of change in the kinds of organisms preserved in the

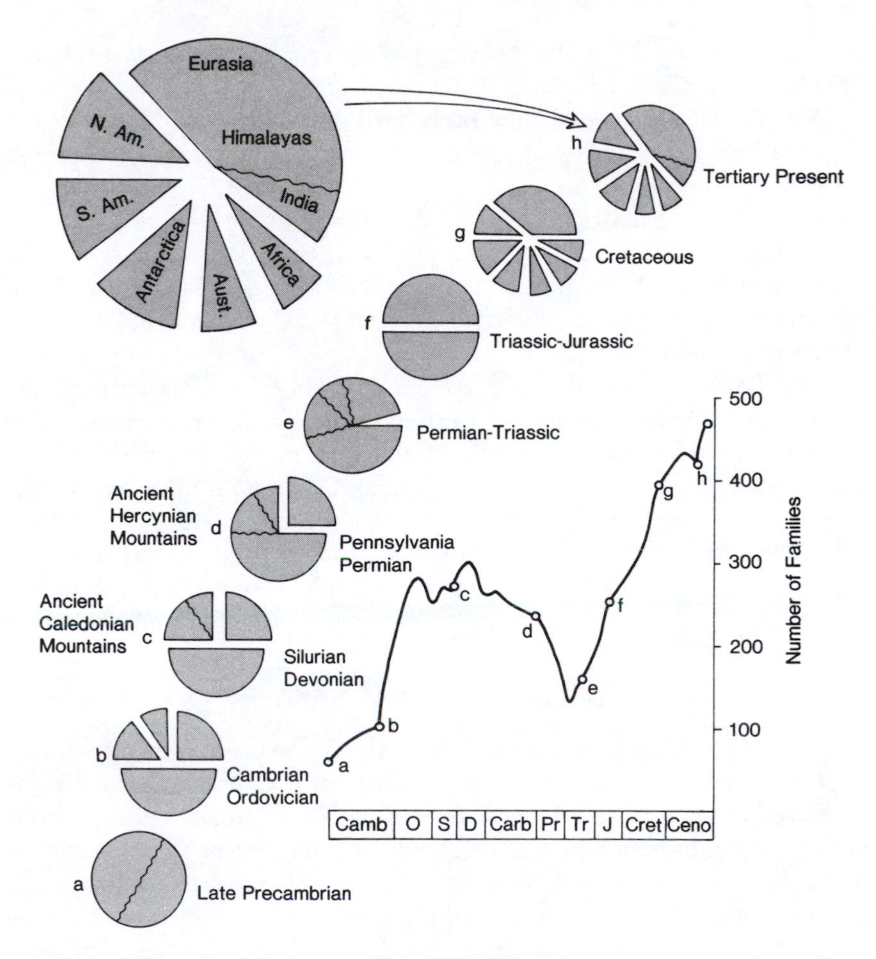

Figure 10.1. Diagram showing continental construction and break-up, and correlation of the diversity of life on earth from Cambrian to Recent time. (After Valentine, J. W., and Moores, E. M., 1972, Jour. Geol., 80(2):167-184)

fossil record. The comparison illustrates the dependency of life on earth to the existing physical environment.

No geologic period is represented less by marine deposits in North America than the Triassic. Characteristic Triassic rocks are red sandstones, mudstones, and shales, and the scarcity of fossils in them poses age assignment and correlation problems of some magnitude. Nonetheless, by using those fossils present and by interpreting paleoenvironmental conditions from sedimentary structures, such as cross-bedding in sandstones and mudcracks and animal tracks in shales, we have been able to restore a picture of Triassic deposition that is predominantly continental in origin.

The Triassic is divided into Early, Middle, and Late. If we do not mention Middle Triassic much in the following discussion, it is because rocks of this age are apparently lacking (or not identifiable as such) in most of North America except in the western eugeosyncline and in Arctic Canada. Since Early Triassic rocks are distributed in somewhat different patterns than the Late Triassic, they have been shown on separate maps, figures 10.2 and 10.3, to accentuate the changes that occurred within this period.

The Triassic Period was first named in 1834 by a German geologist for a three-part sequence of rocks exposed near Hanover, Germany. The rocks he studied are characteristic of the Triassic System worldwide in being mostly continental in origin. To expand Triassic studies worldwide, there is need for a marine fossiliferous sequence to serve as a single world standard. Such rocks exist in the Austrian Alps, and it is there that we base Triassic correlations using the rapidly evolving and abundant ammonoid fossils as the means for recognizing brief time intervals.

BREAK-UP OF PANGAEA

When a continental mass is fragmented, two geologic events commonly, if not universally occur along the rift zone, or line of separation. First is an episode of normal faulting due to tension. This is followed by igneous activity in the form of basalt flows along the fault system and intrusion of dikes into adjacent rocks. Both of these events briefly precede the actual break-up of the land mass. As the rifting develops, an elongate depression forms that may eventually become flooded with sea water and receive marine sediment.

Evidence of the initial rifting of North America from Africa is found in a series of fault bound basins extending in an interrupted belt from the Carolinas to Nova Scotia (fig. 10.4). These fault basins underlie the Connecticut Valley from northern Massachusetts to New Haven, Connecticut, and the New Jersey lowlands from industrial Newark southwestward to Gettysburg, Pennsylvania. These rocks are mostly reddish-brown shales and sandstones, interbedded with basalt flows and sills and are cut by dikes. The igneous rocks have yielded radiometric dates

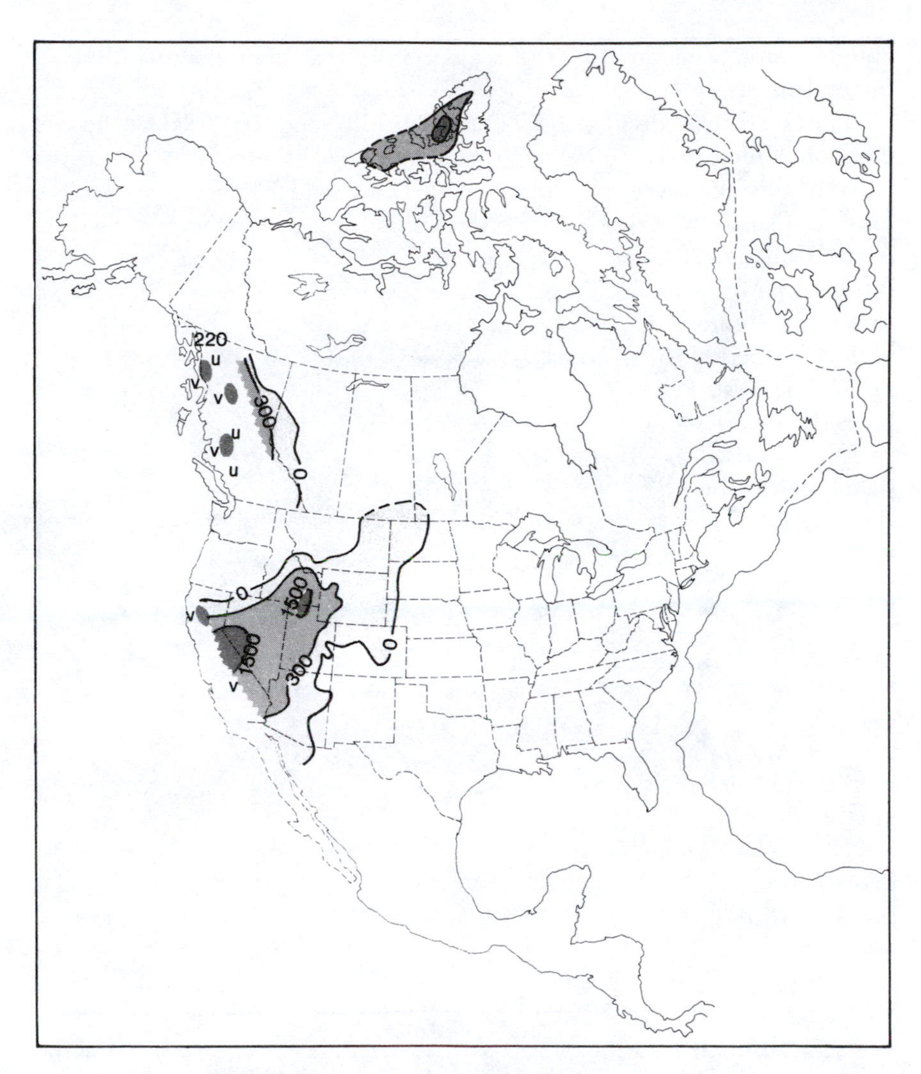

Figure 10.2. Thickness map of Early and Middle Triassic rocks in North America. Thickness in meters. "U" represents ultrabasic intrusions in British Columbia. "V" indicates volcanic rocks associated with cherts and argillites. A Triassic intrusion in the southern Yukon has a radiometric date of 220 million years.

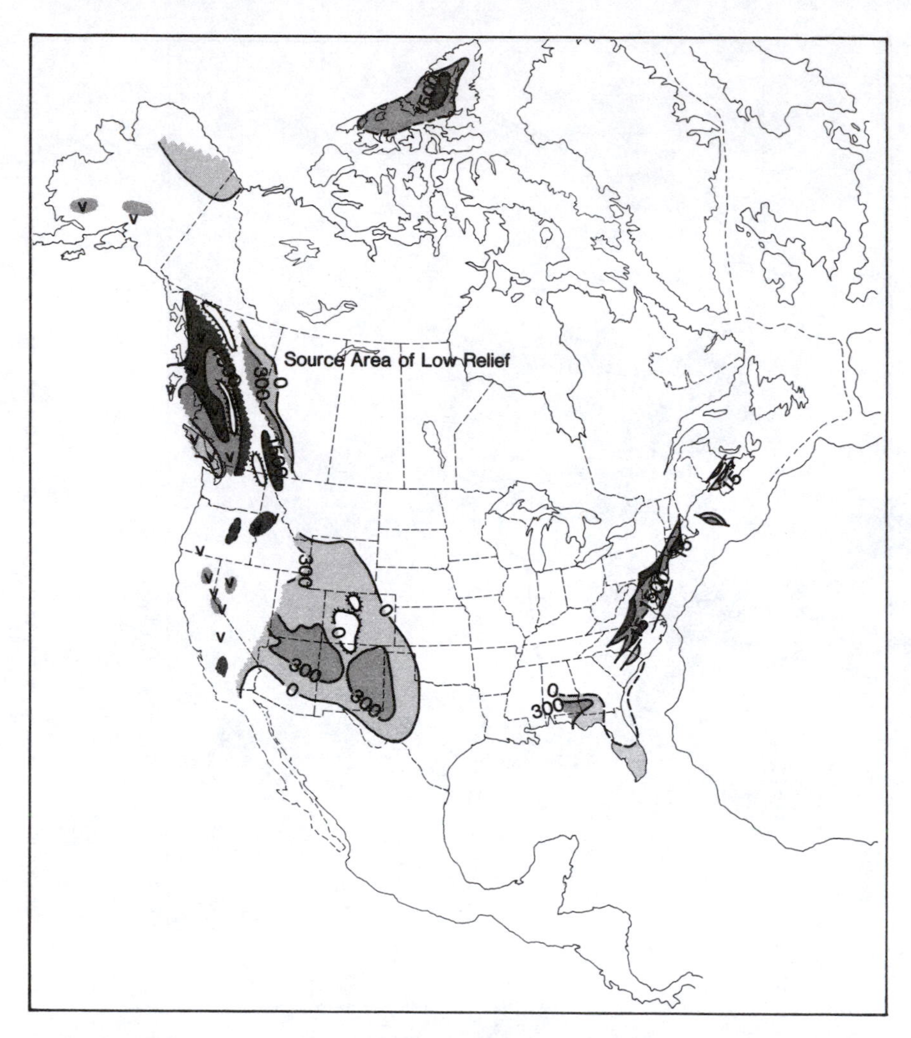

Figure 10.3. Thickness map of Late Triassic rocks in North America. Thickness in metres. "V" represents volcanic rocks. Although Africa began separating from North America during Late Triassic time, Europe did not begin moving away until later Mesozoic time.

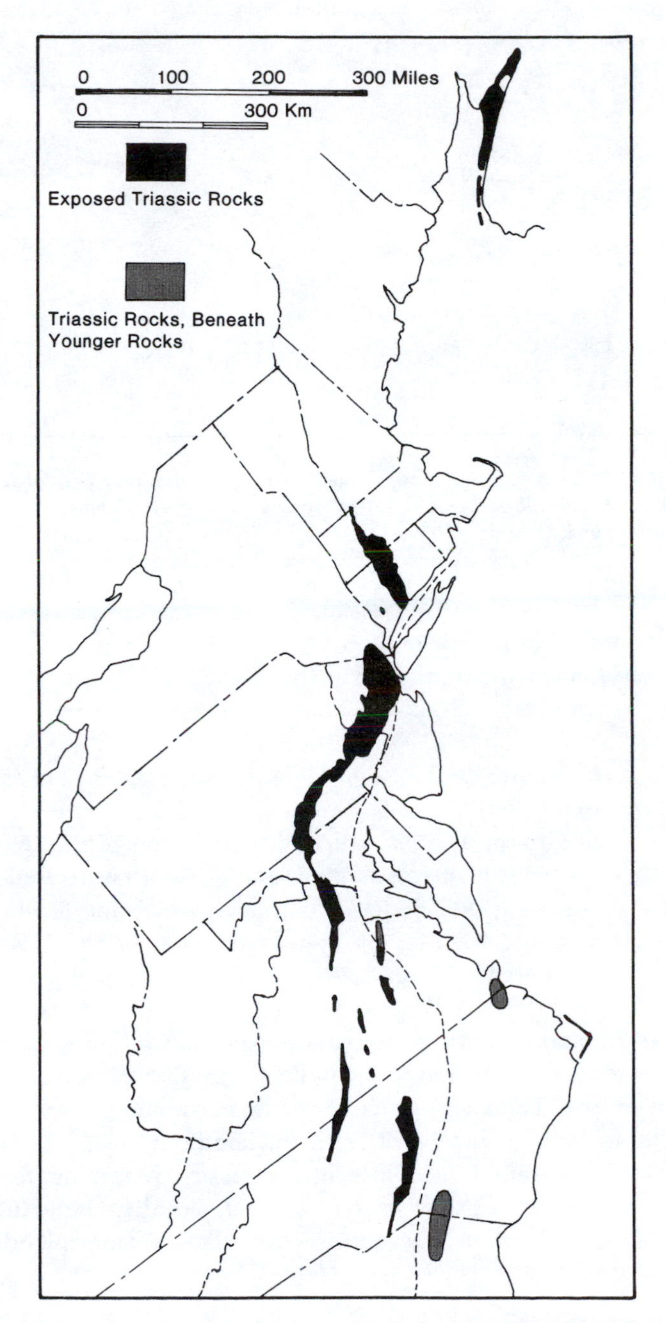

Figure 10.4. Map showing Upper Triassic fault basins of eastern North America.

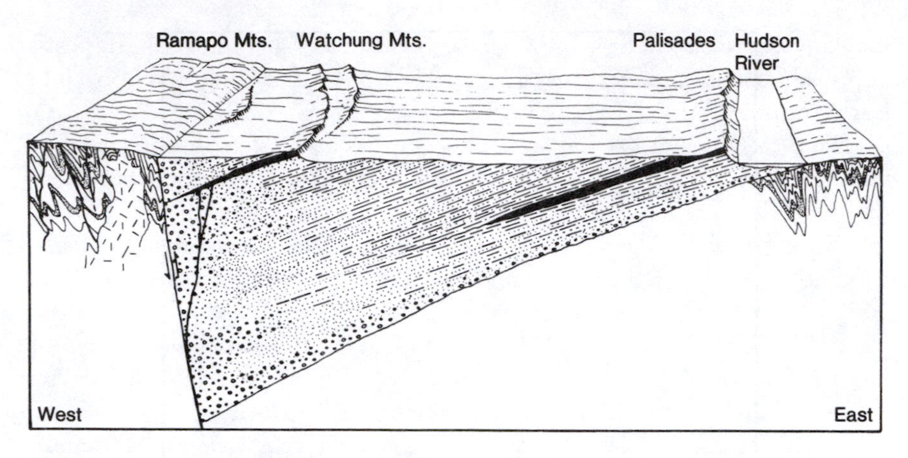

Figure 10.5. Diagram of Upper Triassic strata of the Newark Group deposited in a fault-block basin in New Jersey just west of New York City. Basic lavas and sills (vertical lines) now form ridges. Earlier intense deformation of pre-fault bedrock is shown schematically on either side.

of approximately 190 to 200 million years, or Late Triassic. The basalts are more resistant to weathering than the associated sediments and form some famous features, such as the Palisades along the west side of the lower Hudson River (fig. 10.5), Little Round Top and Round Top of Gettysburg battle fame, the Meriden Hills in Connecticut, and the Holyoke Range in Massachusetts.

The Newark Group, as it is called, becomes conspicuously coarser, sometimes conglomeratic, on the faulted side of each basin. Red Triassic mottled conglomerate is quarried as "Potomac marble" and has been used in numerous buildings, for example, the lower floor of the U.S. Capitol Building. The sediments accumulated in elongate normal fault basins. Erosion of an uplifted block produced sediments that were deposited within the adjoining downfaulted basin, shown in figure 10.5. Fresh feldspars in some of the standstones indicate rapid erosion from a nearly crystalline source. Thickness ranges from a maximum 10,000 metres in Virginia, to an average 2,000 metres in the southern basins.

The Newark Group is a continental deposit, containing fossil land plants and freshwater fish, whose remains are locally abundant in the black mudstones. Dinosaur tracks in the redbeds are more plentiful here than anywhere else in the world. These fossils date the fault basin deposits as mostly Late Triassic with some deposition extending possibly into Early Jurassic.

Deep-sea drilling off the eastern shore has indicated the presence of additional Triassic fault basins buried beneath younger sediments. These too were developed in the earliest phases of continental rifting. Similar fault basins should be expected, and indeed are present, in Liberia, Africa, making the tensional structures symmetrical about the line of separation.

The oldest marine deposits on the eastern coast of North America are Upper Triassic and Lower Jurassic redbeds and evaporites, indicative of the invasion of sea water into the newly opened rift. The high heat flow along a rift zone and the availability of great quantities of sea water provide ideal conditions for the precipitation of marine evaporites, mainly sodium chloride. A modern analog for conditions that prevailed along the Atlantic coast during Late Triassic is the modern Red Sea. As Arabia and Africa separate on either side of a modern divergent plate margin we see the tension faults, igneous activity, and marine sedimentation like that present in the Triassic geology of our east coast.

WESTERN INTERIOR

Early Triassic rocks (fig. 10.2) are scattered over a broad belt from Nevada to North Dakota. The pattern of rock type and thickness in this area is reminiscent of the early Paleozoic. Thick sections of marine strata occur in a geosynclinal belt in western Utah, Nevada, and California, and thin sequences of predominantly nonmarine strata occur on the shelf area from Colorado to the Dakotas. Figure 10.8 is a cross section of the transition from miogeocline to craton in the Idaho-Wyoming area. The geocline, as expected, contains the most continuous record of marine deposits. Desert ranges in southwestern Nevada and adjacent California exhibit more than 1,000 metres of not only Early Triassic rocks but also fossiliferous, ammonoid-bearing Middle Triassic strata. Associated volcanics are found only in the westernmost part of the area, the eugeosyncline. Figure 10.7 is a map showing Late Triassic paleogeography in the western United States.

Triassic rocks record a major change in the depositional pattern of geosynclinal sedimentation in western America. Lower Triassic rocks are

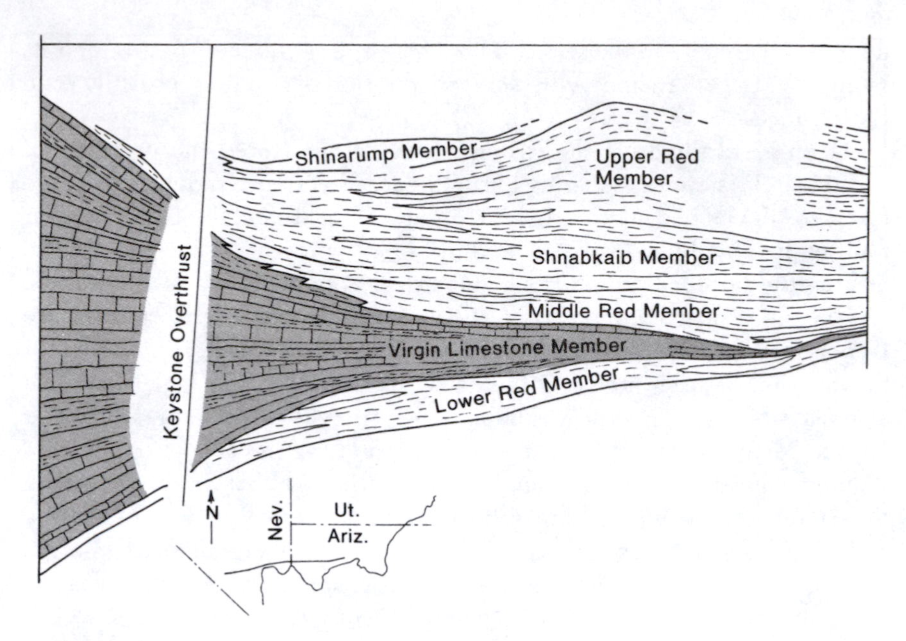

Figure 10.6. Diagram showing Triassic rocks in Utah and Nevada. (After Bissell, H. J., 1969)

the youngest marine rocks in the miogeosyncline. Throughout the Paleozoic, marine sedimentation was by far more common. In the western interior sediments younger than Early Triassic are mostly continental in origin. During Early Triassic time, marine sedimentation occurred in a north-south belt between the craton on the east and the somewhat eroded remnant of the Permian Sonoma Mountains to the west. Lower Triassic rocks (fig. 10.6) form an eastward thinning wedge of limestone and shale, such as seen in the Moenkopi Formation. The Moenkopi Formation is predominantly a mudstone, deposited on a broad flat plain that sloped gently westward across northern Arizona toward the geosynclinal sea in Nevada. Thin tongues of marine limestone indicate three brief times when the seas spread out of the geosyncline across the Moenkopi flats as shown on figure 10.6 Environments interpreted from fossil sedimentary features within the Moenkopi include stream channels, flood plains, fresh- and brackish-water ponds, and shallow seas. Beds of gypsum and casts of salt cubes preserved in the red shales of the Moenkopi indicate widespread aridity. Large reptile tracks and fossils of big, sluggish amphibians in the stream deposits suggest that warm climates

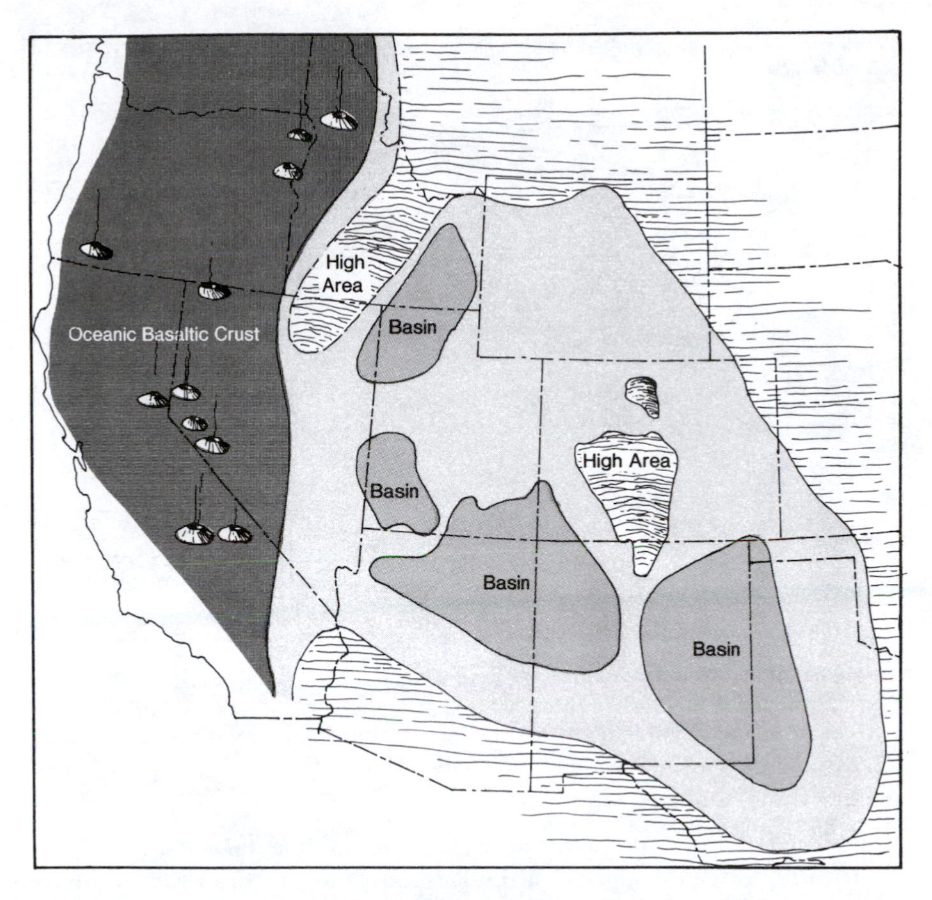

Figure 10.7. Late Triassic basins, highlands, and volcanic islands in western United States. Highlands were source areas for basin filling sediments. (After U.S. Geol. Survey Map 1-300)

prevailed during Moenkopi time. To the north somewhat similar deposits in Wyoming (fig. 10.8) and the Dakotas indicate similar conditions.

Middle Triassic rocks are very restricted throughout North America. They have been identified in western Canada, Nevada, and California on the basis of ammonoids. How much Middle Triassic rock was eroded following the Late Triassic uplift is speculative. Middle Triassic rocks were probably more widespread before that episode of uplift and erosion. In most places in western America, Upper Triassic rocks overlie Lower Triassic strata at a well-marked unconformity. In the Colorado Plateau, for example, the Moenkopi, is overlain by the Upper Triassic Shinarump

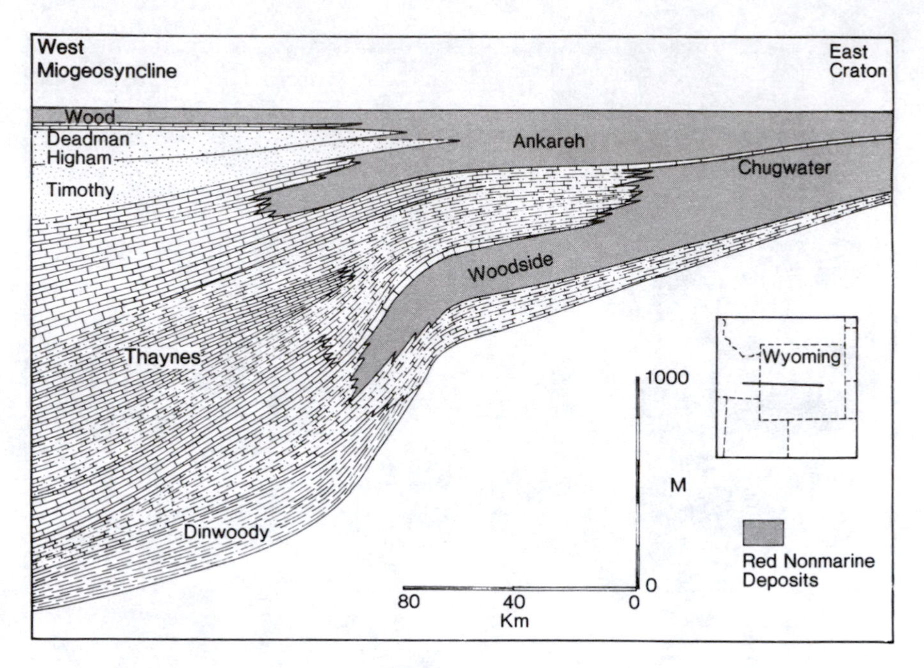

Figure 10.8. Cross section of Triassic strata showing transition from thick, predominantly marine strata of the miogeocline in southeastern Idaho, to thinner predominantly nonmarine cratonic sections in Wyoming. (After Kummel, B., Geol. Soc. Am. Memoir 67, p. 437-468, 1957)

and Chinle Formations, both of which are continental in origin. This relationship is well displayed in the Painted Desert of the Grand Canyon area.

Late Triassic deposits in the western interior are less extensive toward the north than those of the Early Triassic, but extend southward into Texas and New Mexico where they reach a maximum of about 500 metres. A variegated red and gray sandstone and mudstone of continental origin forms scenic badlands between Big Spring and Sweetwater, Texas. Its more famous counterpart in the national parks area of Utah and Arizona is the Chinle Formation. Petrified Forest National Park in Arizona displays thousands of conifer logs that came to rest as driftwood on river floodplains in late Triassic time. This fossil wood is by no means restricted to the Chinle of the Petrified Forest area; it occurs in lesser quantities wherever the formation is exposed in eastern Utah and northern Arizona. Silicification of wood in the Chinle Formation is probably related to the substantial content of volcanic ash that has provided a ready source of silica.

In contrast to early Triassic rocks, Late Triassic rocks are absent from the miogeosynclinal belt in western Utah and eastern Nevada, as shown on figure 10.3. It is possible that they were originally present, but their total absence suggests that they could not have been thick or extensive. Triassic time marks the end of the miogeosynclinal behavior of this area. It ceased subsiding in the Triassic and in early Mesozoic time became a rising mountainous belt, which served as a source for Jurassic and Cretaceous clastic sediments on both flanks.

ALBERTA-NORTHEASTERN BRITISH COLUMBIA

Figures 10.2 and 10.3 show similar thickness trends of Early and Late Triassic strata in western Canada. Triassic rocks are thin on the east, where they occur beneath the Great Plains, and are thicker along the Rocky Mountains, where they are mostly marine siltstones with some interbedded thin sandstones and limestones. Carbonates and evaporites predominate in Upper Triassic strata. Sedimentary features suggest that most of the Triassic sediments were derived from the Canadian Shield and deposited in shallow seas that advanced and receded from the Pacific area numerous times during the Triassic. As in Utah and Nevada, the record of exact relationships between Triassic rocks of the eastern Rockies and Great Plains area and rocks of the west coast eugeoclinal Triassic belt has been obliterated by later Mesozoic tectonism.

WEST COAST EUGEOSYNCLINE

Lower Triassic eugeosynclinal rocks are rare along the west coast compared to Upper Triassic eugeosynclinal deposits. Late Triassic rocks, typically eugeosynclinal assemblages, include greenstone, shale, andesite flows, and volcanic breccia, with occasional chert pebble conglomerates. Several small granite intrusives in northern British Columbia and the southern Yukon yield radiometric dates centering around 220 million years, or early Triassic. Evaluation of sedimentary patterns in Late Triassic time in British Columbia suggest that part of the sediment deposited in the sinking troughs may have been derived from the Canadian Shield to the east, but most was derived from the offshore volcanic islands and tectonic uplifts to the west. One of the most widespread, and thereby most useful, fossils that occurs in Late Triassic shale and siltstone along the west coast is the thin-shelled bivalve *Monotis*, whose coquinas come as a welcome find to geologists searching these rather barren west coast sequences for some clue as to age.

ALASKA-YUKON

The west coast Late Triassic volcanic belt can be extended northward from British Columbia into southern Alaska, where it bends westward. Here *Monotis*-bearing siltstones are interbedded with andesites and greenstones. Several hundred kilometres to the north in the Brooks Range, Alaska, and adjacent mountains in the northern Yukon, Late Triassic rocks are nonvolcanic, *Monotis*-bearing, marine, calcareous sandstones ranging from zero to 300 metres thick.

ARCTIC ISLANDS

Nearly 5,000 metres of Triassic sediments accumulated in the central part of the Svedrup Basin. These deposits are mostly marine siltstone, sandstone, and shale with some conglomerates. Associated fossils indicate that deposition in the central part of the basin was essentially continuous from Late Permian through Late Triassic time. Sediments were derived mainly from uplands to the south and east, as well as occasionally from the north. That a continuous marine connection was probable between the Sverdrup Basin and the Arctic Ocean is indicated by a cosmopolitan marine Triassic fauna preserved in the basin strata.

TRIASSIC LIFE

During the 35-million-year history of the Triassic Period, great changes occurred in the life forms on earth, both plant and animal. Following the massive extinctions at the close of the Paleozoic Era (fig. 10.1), many new life forms evolved as part of the Triassic biosphere. Reptiles were by far the most successful group of organisms to develop in the Triassic, so much so that the entire Mesozoic Era is called "the Age of Reptiles." Molluscs, too, attained new heights of abundance and diversity. The environmental niches commonly occupied by brachiopods during the Paleozoic became filled with bivalves, whose fossil remains are as abundant in Mesozoic and Cenozoic strata as brachiopods are in the Paleozoic beds.

Cephalopods were strongly affected by the environmental changes accompanying the close of the Permian. Goniatite cephalopods, which were the dominant ammonoid of the Late Paleozoic, were replaced in their position of dominance by the ceratite ammonoid. Indeed the entire ammonoid stock came near to extinction at the close of the Paleozoic and again at the end of the Triassic. In each of these cases a single family survived and became the root stock for succeeding generations. The ceratites of the Triassic form the basis of time resolution during the period. Belemnoids, a squidlike form with a heavy external skeleton, left abundant fossils in Triassic rocks and even more in the overlying Jurassic and Cretaceous strata.

Corals were modernized following the Permian extinction of the Paleozoic types, and forms like those that now thrive in warm clear shallow waters of the tropical zone became common. The oldest known Mesozoic corals are Middle Triassic in age, and until Early Triassic forms are found paleontologists will speculate on the evolutionary relationship with their extinct predecessors.

Triassic vertebrates contain many new forms, mainly reptiles, ushering in the Mesozoic "Age of Reptiles." A common Triassic river reptile was the phytosaur, resembling modern crocodiles but with a long slender snout. Some advantage to life on land had been the adaptive pressure behind the origin of reptiles when they evolved from amphibians. Early in the Mesozoic, some reptiles reversed this trend and returned to life in the sea. Marine reptiles include the dolphinlike ichthyosaur and the long-necked plesiosaur. Both fish-eating carnivores, their remains are recovered from marine strata throughout the Mesozoic.

Dinosaurs appeared in the geologic record during the Triassic, and quickly rose to dominate the land for the remainder of the Mesozoic. Triassic dinosaurs were small, less than 5 metres from head to tail, hence are seldom displayed in museums, which usually feature the more spectacular giants of the Jurassic and Cretaceous. Most were bipedal, balancing themselves on their tail. Their three-toed footprints were for a time mistaken for giant bird tracks.

A major discovery of Triassic reptile and amphibian fossils was made in Antarctica in 1978 at a locality some 500 kilometres north of the South Pole. Many of the animals represented in this collection are also found in contemporary rocks in Africa, Australia, and India. This is another bit of evidence to support paleogeographic reconstructions of the Triassic land mass, Pangaea.

From the mammallike reptiles, or therapsids, evolved the first true mammals. From Upper Triassic rocks in Wales, well-differentiated teeth and jawbones clearly belonging to a mammal have been collected. Mammals differ from their reptile ancestors in several ways: differentiated teeth modified to accommodate a varied diet, a single-boned jaw, and an enlarged brain case. Early mammals were small inconspicuous animals among their contemporaries and remained so until the great predatory reptiles, the dinosaurs, became extinct at the close of the Mesozoic Era.

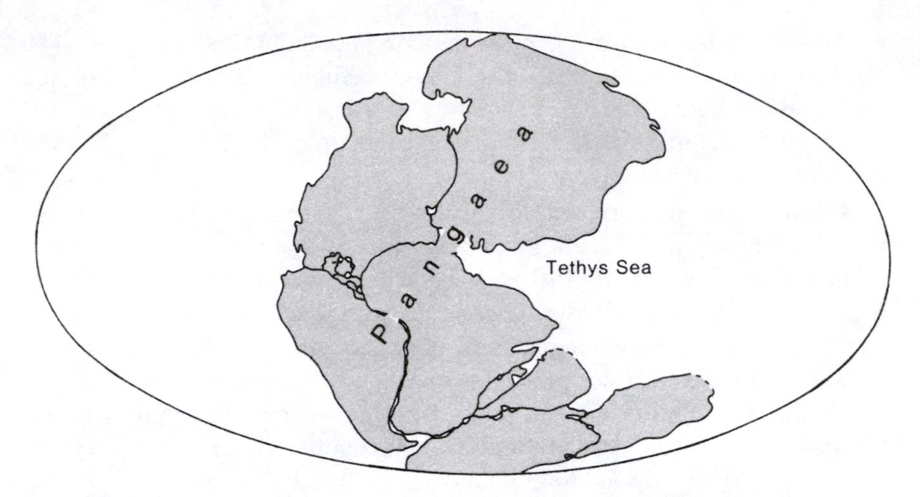

Figure 10.9. Diagram showing the ancient landmass Pangaea approximately 200 million years ago. (After Dietz, R. S. and Holden, J. C., Scientific American, October 1970)

In the Newark Group, plant fossils include leaves of ferns, scouring rushes, conifers, and cycads. The Petrified Forest in Arizona preserves logs of conifers mainly, but leaves of cycads and ferns are also present within the Chinle Formation. Triassic forests were dominantly pine, a sharp contrast to the tropical types characterizing the Pennsylvanian. The last occurrence of the great scale tree, *Sigillaria,* so abundant during the Pennsylvanian, is found in Triassic rocks. Except for these occurences, Triassic land plants are not widespread as fossils, perhaps due to the arid conditions that characterized the period.

TRIASSIC TECTONICS

The oldest rocks on the floors of the world's oceans are approximately 180 million years old, or Early Jurrasic. Structural and paleomagnetic studies indicate incipient separation of North America from Africa and South America late in the Triassic. Most Triassic rocks record the existence of Pangaea as shown in figure 10.9. Compare 10.9 and 10.10, which show the initial separation of Laurasia and Gondwana and the incipient development of the Atlantic Ocean along a divergent plate margin, or spreading center. The Triassic fault basins discussed earlier in this chapter, as well as the presence of Upper Triassic evaporites and related deposits along the eastern margin of North America, attest to this split-

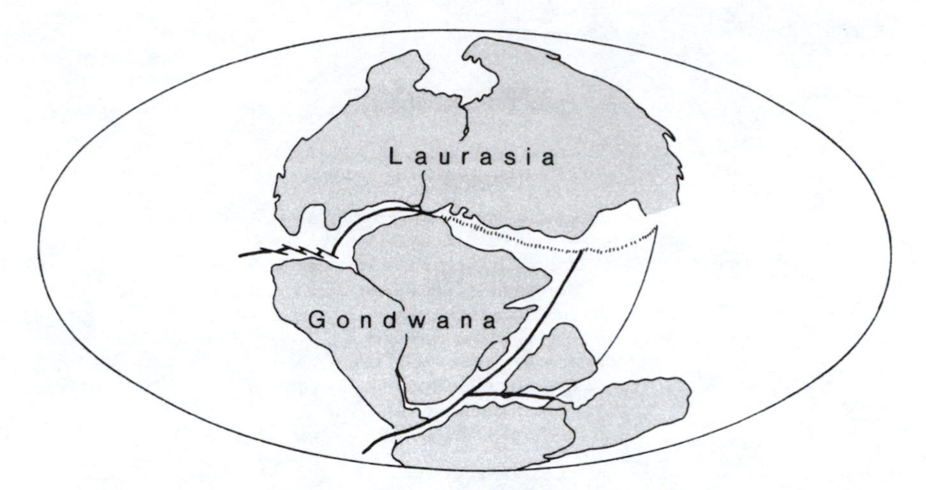

Diagram 10.10 Diagram showing initial break-up of Pangaea at the close of the Triassic Period, approximately 180 million years age. (After Dietz, R. S. and Holden, J. C., Scientific American, October 1970)

ting. From this point in the Late Triassic, the major physical evolution in the distribution of continents and ocean basins has been the continuing break up of Pangaea. Before the end of the Mesozoic, North America and South America had completely separated from Gondwana, and except for a connection between Arctic Canada and Europe, they had also split away from Laurasia. Pangaea was finally split into our modern continental masses in the Early Tertiary, some 45 million years ago.

11

Jurassic

TOPICS

Early Jurassic of the Colorado Plateau
Jurassic Eugeosynclinal Deposits
Upper Jurassic—Upper Cretaceous Franciscan Mélange
Sundance—Zuni Sea
Morrison Formation
Gulf Coast Jurassic
Jurassic of the Arctic Coast
Nevadian Orogeny
Jurassic Life
Jurassic Tectonics

Jurassic rocks of North America are exposed in the western interior, along the entire Pacific coast, and in the Arctic coast of Canada (fig. 11.1). Deep drilling in the Atlantic and Gulf Coastal Plains has revealed Jurassic sediment that contains abundant evaporites. Throughout the Rocky Mountains and Colorado Plateau Jurassic strata are exposed around most of the uplifts and form much of the classic scenery of these areas.

Jurassic rocks record a major invasion of the sea from the north, as well as the early development of the Gulf of Mexico. These two seaways were apparently only occasionally, if ever, connected.

With the new pattern of motion of the continent, as North America began its westward travel early in the Jurassic, the Pacific Coast, now the leading edge of the continent, underwent a massive alteration as the Sierra Nevada Mountains began to be uplifted along the convergent margin of the North American Plate.

EARLY JURASSIC OF THE COLORADO PLATEAU

Early Jurassic rocks of the Colorado Plateau represent continental deposits, which accumulated along the southwestern margin of the craton (fig. 11.2). The thickest and most prominent of the Lower Jurassic formations on the Colorado Plateau is the Navajo Sandstone, a promi-

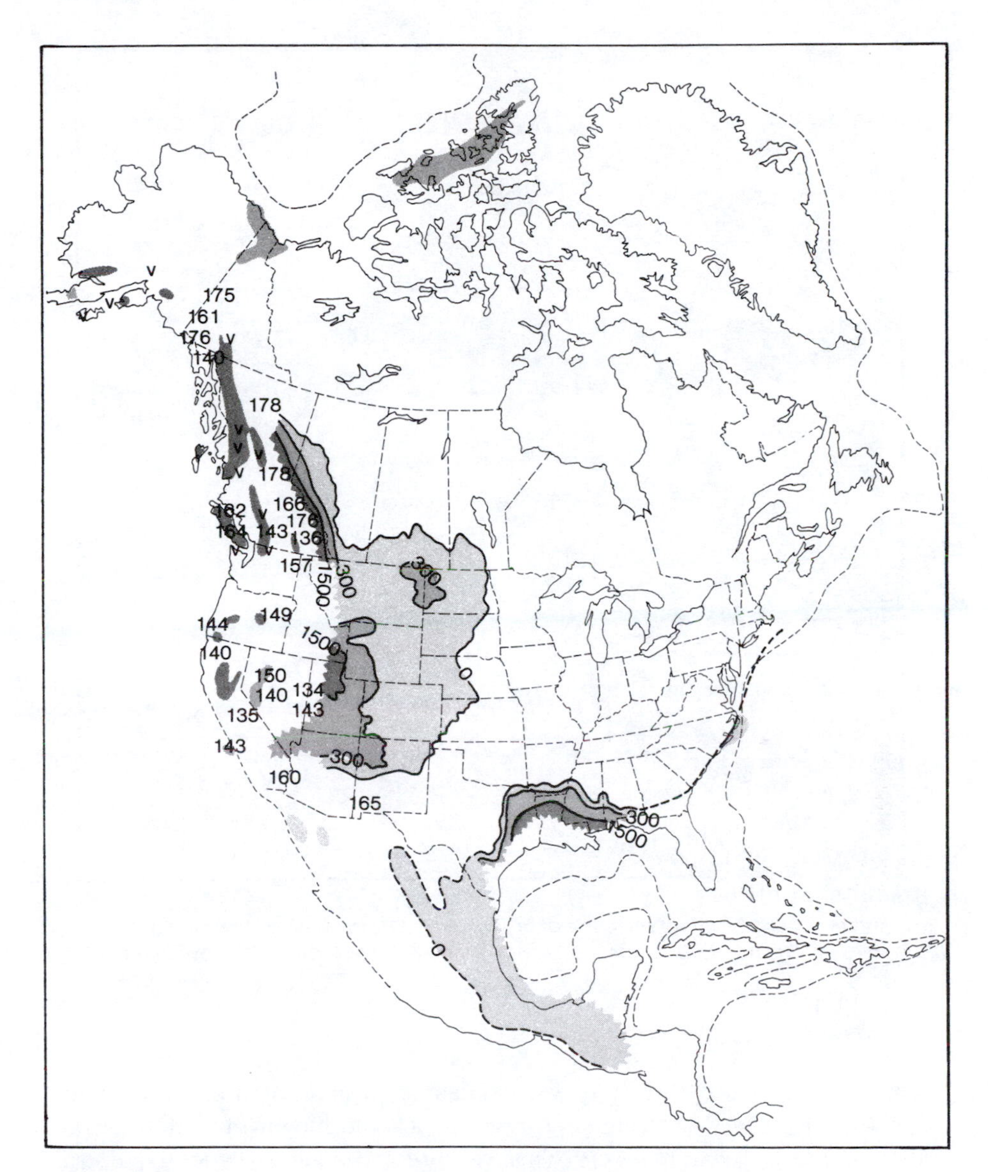

Figure 11.1. Thickness map of Jurassic System in North America. (Thickness figures in metres with thickest deposits shown by heaviest shades: "V" indicates occurrence of volcanic rocks. Radiometric dates for intrusive rocks shown in millions of years.

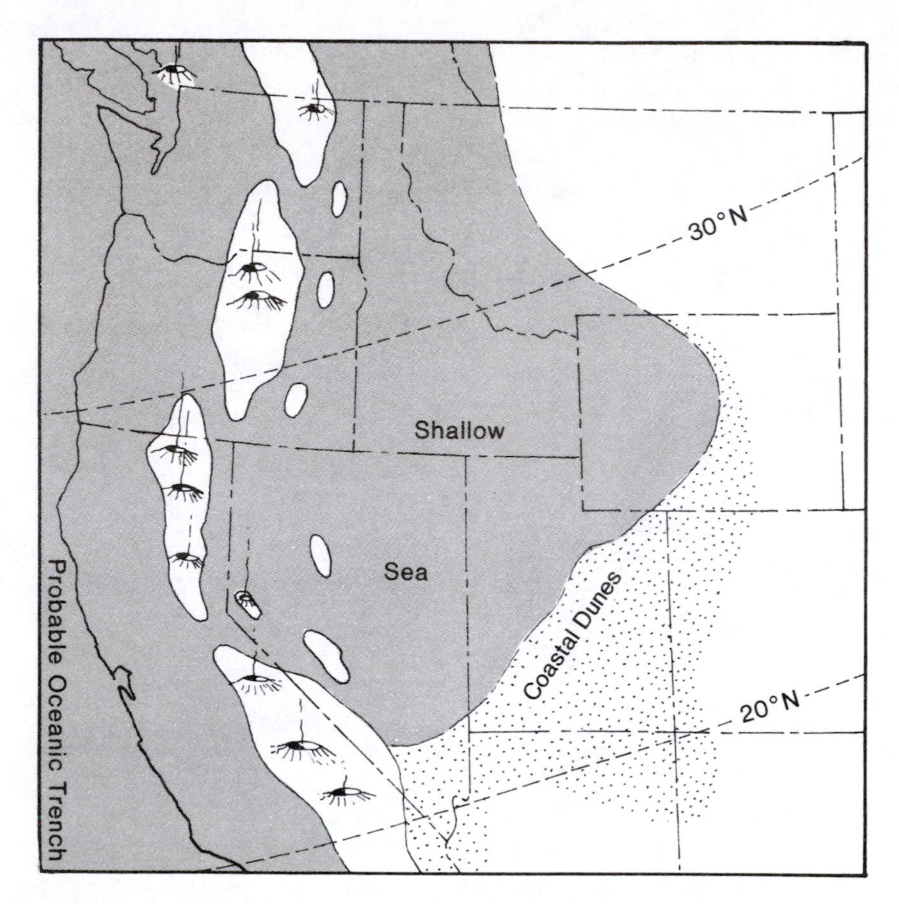

Figure 11.2. Paleographic map of western North America showing area covered by inland Jurassic sea and marginal coastal dunes. (From Stanley, K. O., Jordan, W. M. and Dott, R. H. Jr., Am. Assoc. Petrol. Geol. bull. v. 55, p. 13. 1971)

nently cross-bedded, white quartz sandstone, which outcrops throughout the Plateau. The Navajo is interpreted as a wind-blown sand deposit on a widespread Jurassic desert. The well-preserved, cross-bedded pattern represents stratification on the lee side of numerous sand dunes built up successively to a thickness of over 300 metres. In recent years a few geologists have suggested a marine origin for at least part of the Navajo. Cross-bedding is interpreted by them as typical of shallow underwater dunes such as those found in modern marine environments. Recently dinosaur and crocodile fossils were described from near the top of the Navajo Sandstone in northern Arizona. The presence of such fossils seems incompatible with a desert environment, and on the basis of these kinds of data perhaps the uppermost Navajo represents transitional deposits between unquestionably wind-blown deposits below and at least locally aqueous deposits above. Disconformably overlying the Navajo are marine deposits that would fit this general picture of transition.

JURASSIC EUGEOSYNCLINAL DEPOSITS

From California to Alaska, great thicknesses of volcanic rocks and typical eugeosynclinal sediments accumulated. Up to 6 kilometres of impure sandstone, dark slates, and chert conglomerates are interbedded with basalt and volcanic breccia. Much of the basalt is pillow basalt indicating submarine eruptions. In central British Columbia, over 6 kilometres of monotonously uniform volcanic breccia and basalt exist. These volcanic-rich deposits represent an extremely active period of volcanism along the leading edge of the westward moving American plate. Overlying these rocks in the eugeosyncline are a series of dark shales and slates over 1.5 kilometres in thickness. These younger eugeosynclinal sediments locally contain Jurassic ammonoids, belemnoids, and bivalves, indicating a change in the sedimentary pattern within the sinking eugeosyncline. Sediments and volcanics in the eugeosyncline were derived from the erosion of volcanic islands west of the present Pacific coastline. A small percentage of the sediments apparently were transported from the craton to the east.

UPPER JURASSIC—UPPER CRETACEOUS FRANCISCAN MÉLANGE

Late Jurassic to Late Cretaceous rocks in the Coast Ranges of California consist of a chaotic mixture of repeating rock types, including deepwater graywacke, siltstone, and black shale. Blocks of metamorphosed rocks from the Franciscan Formation are included, and the whole mixture is interbedded with volcanics amounting to approximately 10 percent, mostly pillow basalts and volcanic breccia. The mélange rests directly upon oceanic crust – the continental crust is missing. Lenses of chert and shale contain abundant radiolarians, or siliceous protozoans, whose shells are common today in oozes on the deep ocean floor. Fossils of benthic or bottom-dwelling organisms are absent.

East of the Franciscan rocks, separated by a major thrust fault within the Great Valley of California, rocks of the same age consist mostly of

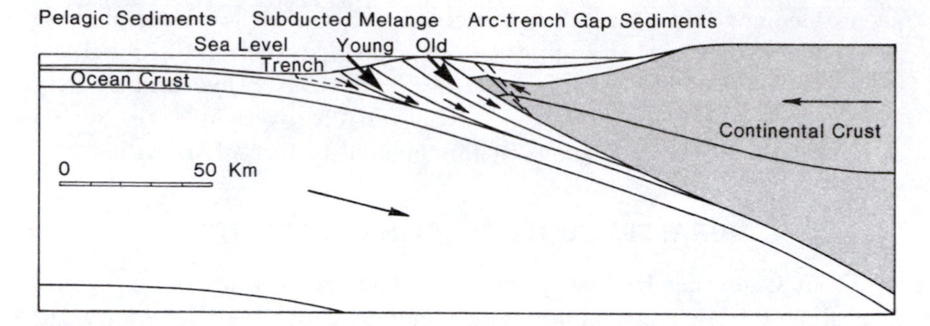

Figure 11.3. Cross section showing convergent plate margin and development of melange similar to conditions which existed along the Pacific coast of North America during Jurassic and Cretaceous times. (After Windley, B. F., Evolving Continents, p. 253, John Wiley and Sons, 1977)

sandstone, siltstone, and shale with considerable conglomerate. Benthonic fossils are abundant in these strata.

The Franciscan sequence of Jurassic and Cretaceous rocks is interpreted to represent oceanic basin, continental slope, and continental shelf sediments brought together as North America overrode the subducting Pacific Ocean floor (fig. 11.3). The Franciscan mélange represents sediments from all parts of the deep ocean environment, midocean ridge, ocean floor, island arc, and trench mixed with pieces of the oceanic crust and upper mantle.

Deep ocean materials were scraped off against the continent as the oceanic plate slid into a subduction zone. The Great Valley sequence represents miogeosyncline deposition on the continental slope and shelf, with abundant benthic fossils, typical clean sedimentary rock types, and no volcanics.

SUNDANCE-ZUNI SEA

Early in the Jurassic, the Sundance-Zuni seaway encroached on the western interior of the continent from the north (fig. 11.4). The sediment deposited in the epicontinental sea represents four separate transgressive-regressive cycles of the sea upon the craton. Source areas for these would include the craton to the east, volcanic highlands to the west, and local uplifts within the sea itself (fig. 11.4). Molluscan fossils are particularly abundant in these deposits attesting to the shallow, highly

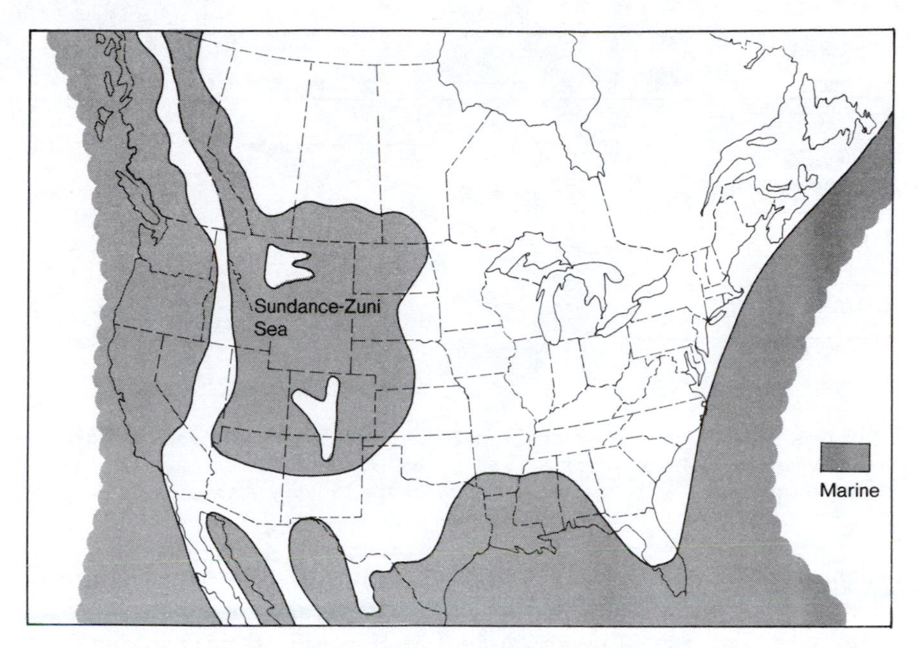

Figure 11.4. Paleogeography of Late Jurassic time in North America. (After U.S. Geological Survey Map I-175)

favorable marine conditions that existed in the sea. Near the town of Fernie, British Columbia, in the marine Fernie Shale, is preserved perhaps the largest of all known ammonoids. Still in its enclosing rock on the mountainside, the conch measures two metres across.

Apparently the Sundance-Zuni Sea was never able to connect with its counterpart to the south as the Gulf of Mexico began to flood the southern margin of the craton (fig. 11.4). There is at least some indication that the Sundance-Zuni Sea did carry over to the geosyncline to the west, as illustrated on figure 11.2.

MORRISON FORMATION

The youngest Jurassic deposit in the western interior of the United States is called the Morrison Formation (fig. 11.5), and overlies deposits of the Sundance-Zuni Sea. The Morrison Formation is named for a small town near Denver, Colorado; these late Jurassic deposits are among the most interesting on the continent. The Morrison averages only 130 metres in thickness, but is spread over an area in excess of 1.3 million square

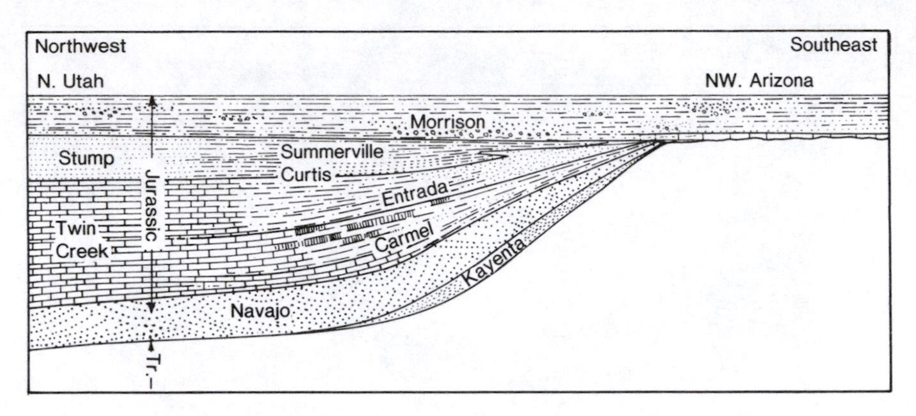

Figure 11.5. Cross section of Upper Triassic and Jurassic strata in the Colorado Plateau area. Erosion of these rocks has produced most of the landforms which make up classic "western" scenery of the United States.

kilometres (fig. 11.6), where it displays its unique color blend of green, gray, and maroon. Interbedded shales, siltstones, and conglomerates were accumulated on a great floodplain by Late Jurassic streams carrying their heavy load of sediments eastward from the uplifted engeoclinal area. Morrison sediments record Late Jurassic mountain building. Sediments of the Morrison were derived from mountains forming in the eugeosyncline to the west. Streams flowing eastward deposited their loads of sand, silt, clay, and gravel on a wide floodplain upon which roamed the great dinosaurs.

Preserved within the Morrison strata is the world's richest known storehouse of Jurassic dinosaur fossils. Included in almost every major museum in the western world are huge fossilized bones of Morrision dinosaurs. *Brontosaurus, Allosaurus, Brachiosaurus, Stegosaurus,* and many more have been excavated from these rocks. Dinosaur National Monument in eastern Utah displays a massive outcrop of Morrison sandstone literally filled with various bones of dinosaurs. The Morrison Formation is also one of the most important producers of uranium. Many of the largest occurrences of the metal are in stream channels within the formation.

The age of the Morrison was unsolved for many years because of the lack of marine invertebrate fossils that are used to subdivide Jurrasic time. Not until Upper Jurassic ammonoids were found in Tanzania, East Africa, in strata which were interbedded with nonmarine strata containing Morrison-like dinosaur fossils, was the age firmly established as latest Jurassic.

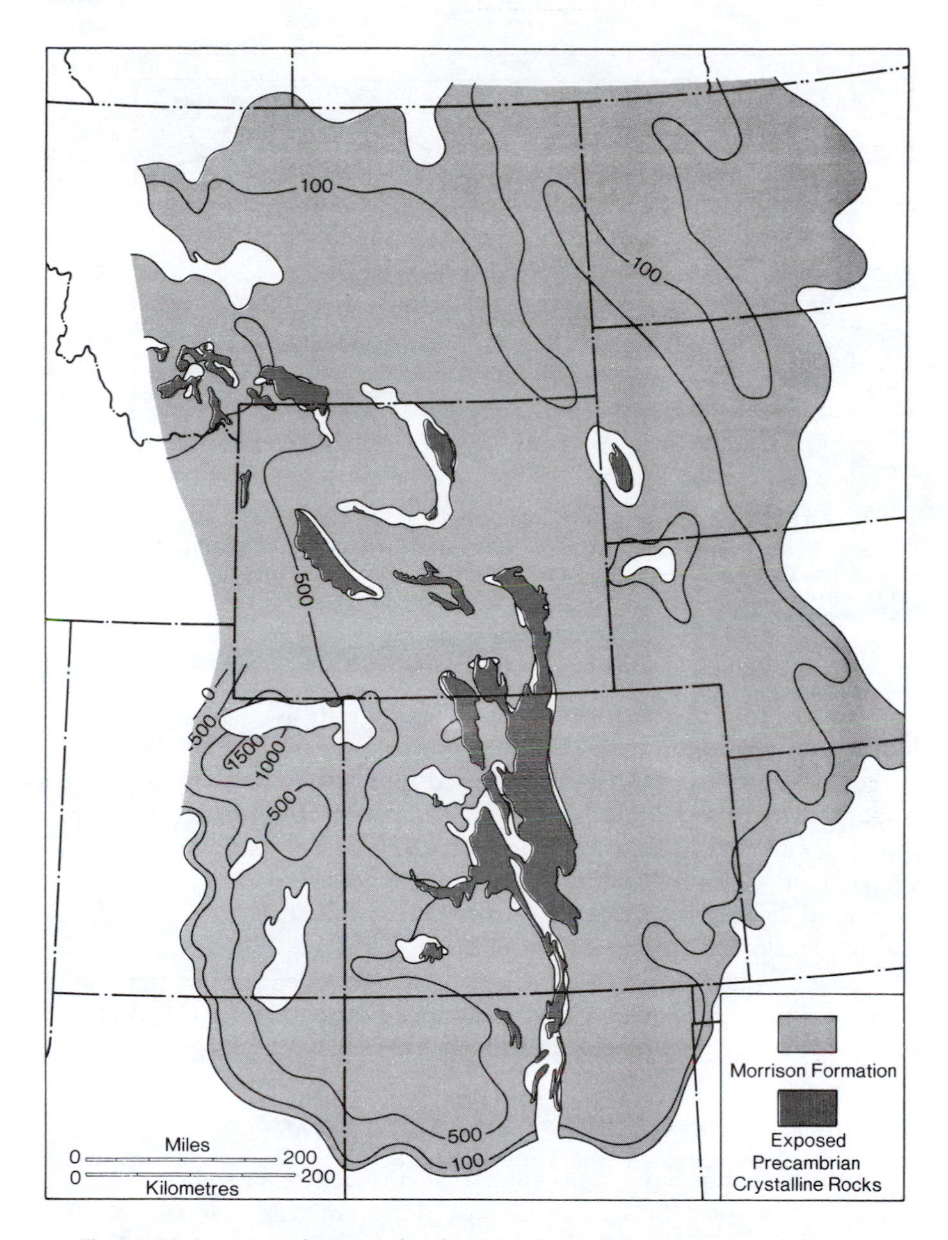

Figure 11.6. Map showing distribution of Morrison Formation of Jurassic age. (After Peterson, J. A., 1972, p. 185, Geol. Atlas of Rocky Mtn. Region, Rocky Mtn. Assoc. of Geol.)

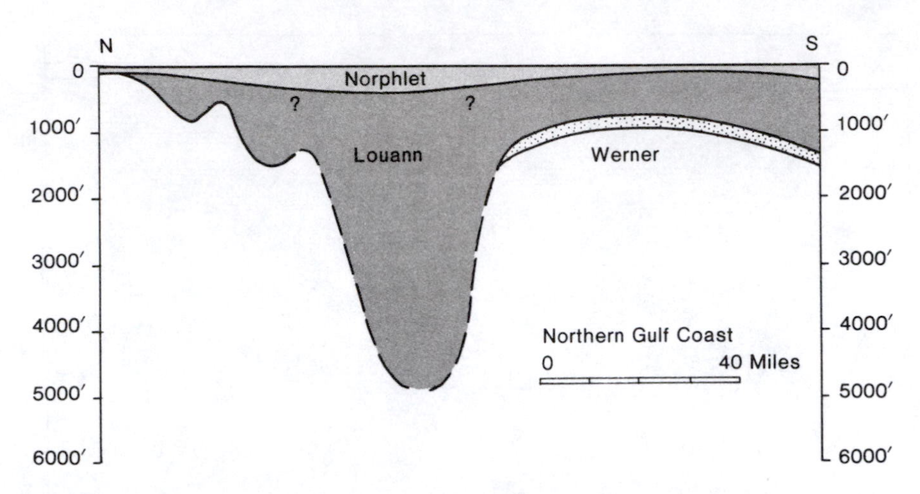

Figure 11.7. Cross section showing Middle Jurassic strata in northern Gulf Coast. (From Shell Strat. Atlas of North and Central Amer., Cook, T. D. and Bally, A. W. editors, 1975, p. 187)

GULF COAST JURASSIC

Middle Jurassic evaporite deposits, mainly salt, are widespread and locally over one and a half kilometres in thickness (fig. 11.7). This salt, called the Louann Salt, was deposited during the initial rifting of the Atlantic Ocean early in the Jurassic. The climate of the Jurassic in North America was very likely warm and dry, ideal conditions for widespread evaporation. The salt of the Jurassic has immense economic interest because it has risen in domelike structures within the overlying strata and has localized vast quantities of oil (fig. 13.6).

Above the Louann Salt is an Upper Jurassic sequence of marine sediment, including limestone, shale, and more evaporites (fig. 11.8). It is into these beds, and others, that the salt domes have intruded.

JURASSIC OF THE ARCTIC COAST

In the Sverdrup Basin along the Canadian Arctic Coast, deposition of sediment was more or less continuous throughout the Jurassic. Sandstones and shales predominate, and total about one kilometre in thickness. Most of the sediment was derived from the craton to the southeast (fig. 11.9). Westward along the Alaskan North Slope Jurassic strata are much thicker, totaling over four kilometres of deep water sediment called turbidite, with shallower water shales and sandstones.

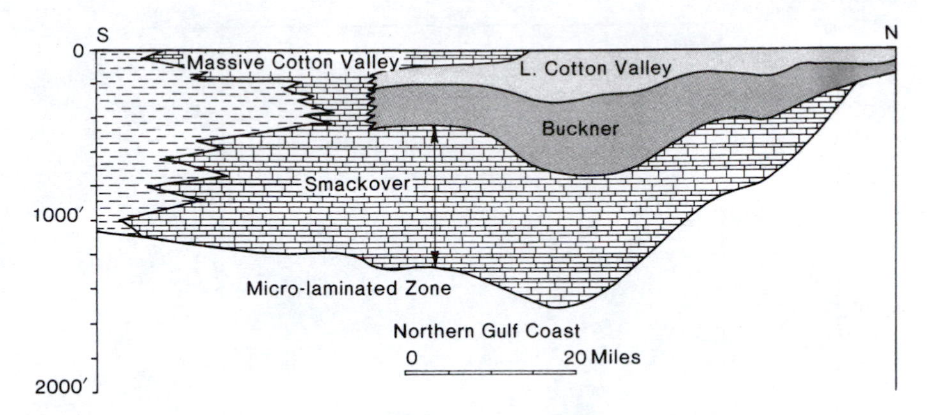

Figure 11.8. Cross section showing Upper Jurassic strata in northern Gulf Coast. (From Shell Strat. Atlas of North and Central Amer., Cook, T. D. and Bally, A. W. editors, 1975, p. 191)

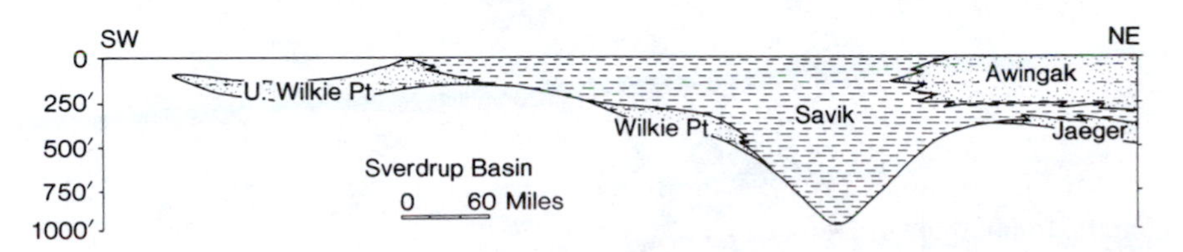

Figure 11.9. Cross section showing Upper Jurassic strata in northern Gulf Coast. (From Shell Strat. Atlas of North and Central Amer., Cook, T. D. and Bally, A. W. editors, 1975, p. 191)

NEVADIAN OROGENY

A major episode of mountain building marks the end of the Jurassic Period. From Baja California to Alaska, enormous quantities of granite and granitelike rocks were intruded into the eugeosyncline or formed by melting of eugeosynclinal rocks (fig. 11.10). Radiometric dates from the granites reveal a complex array of numerous batholiths, whose ages range from 140 to 80 million years. This was only the first stage of a continuous wave of deformation along the western margin of the continent. Beginning 140 million years ago during the Jurassic, and ending 80 million years ago, the Nevadian Orogeny represents the first major pulse of tectonism during the Mesozoic along the western margin of the continent. The present Sierra Nevada, Klamath, and Coast Range Mountains, as well as the granititic rocks of Baja, California, are remnants of the Nevadian Orogeny. While the wave of deformation moved inland,

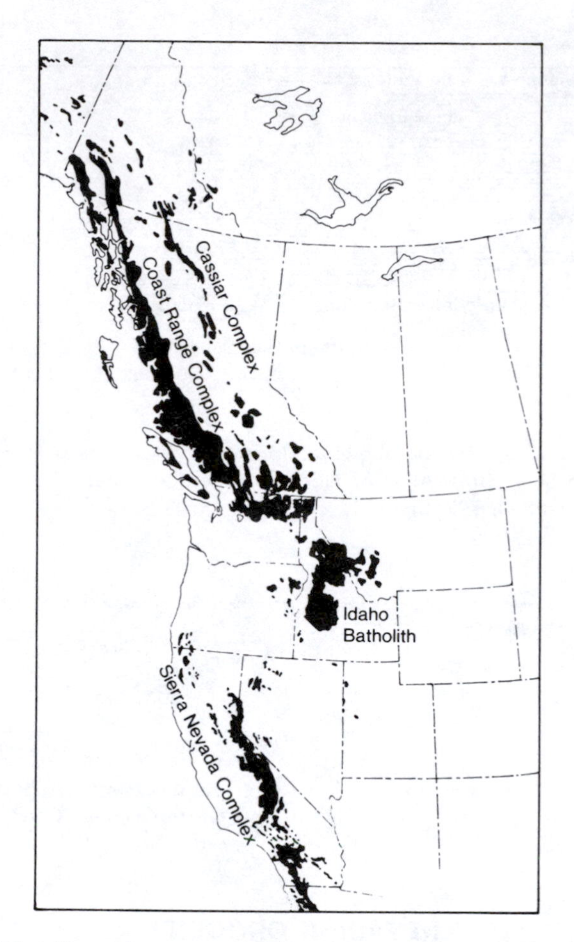

Figure 11.10. Map of western North America showing the distribution of Jurassic and Cretaceous granitic intrusions. These intrusions form a complex ranging in age based on radiometric dating from 80-140 million years.

continued igneous activity took place within the eugeosyncline, emplacing younger intrusive masses until the compression ceased at the 80-million-year mark.

Some 200 separate intrusions can be identified in the Sierra Nevada alone. The intrusions are thought to have been emplaced at depths of from six to ten kilometres below the surface, largely beneath a cover of their own volcanic ejecta.

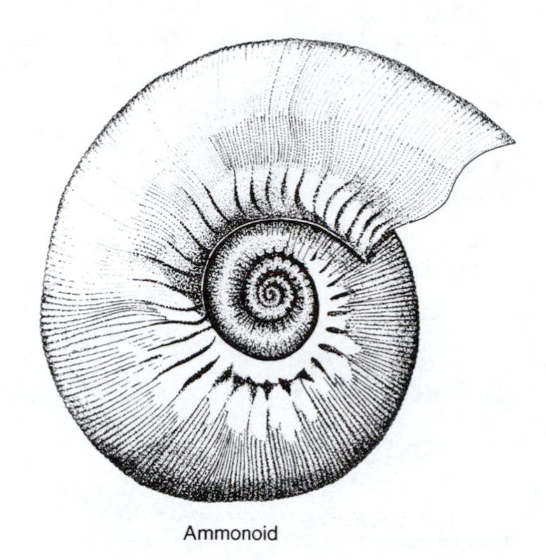

Figure 11.11. A typical Jurassic ammonoid cephalopod.

JURASSIC LIFE

Marine invertebrates, again dominated by molluscs, ammonoids, bivalves, and belemnoids especially, flourished in the sea. Jurassic ammonoids (fig. 11.11) reach their apex as stratigraphic indicators, resolving increments of approximately one million years. Other molluscs, bivalves, and gastropods, are abundant and widespread as fossils in marine deposits. Crinoids and echinoids represented the most successful types of echinoderms.

In a fine, soft shale in southern Germany the first fossil bird appears in the geologic record. Resembling its reptilian cousins from which it had only recently evolved, the first bird was about the size of a crow. The imprint of feathers is clearly visible on each of the two available specimens. Perhaps an even older specimen has recently been discovered on the Colorado Plateau.

Life on land was totally dominated by dinosaurs. Some Jurassic dinosaurs grew to lengths up to 28 metres. The largest forms were herbivorous. Among the best known Jurassic dinosaurs are *Brontosaurus, Diplodocus, Stegosaurus* (fig. 11.12), and *Allosaurus.*

Ichthyosaurs and plesiosaurs (fig. 11.13), which measure up to three and five metres respectively in length, were both common in Jurassic seas, much more so than in the previous period.

Stegosaurus

Figure 11.12. **Stegosaurus** is a dinosaur collected from the Morrison Formation, first discovered at Como Bluff, Wyoming. The plates and spiked tail were a defensive adaptation against his predacious contemporaries. He stood about 3 metres high.

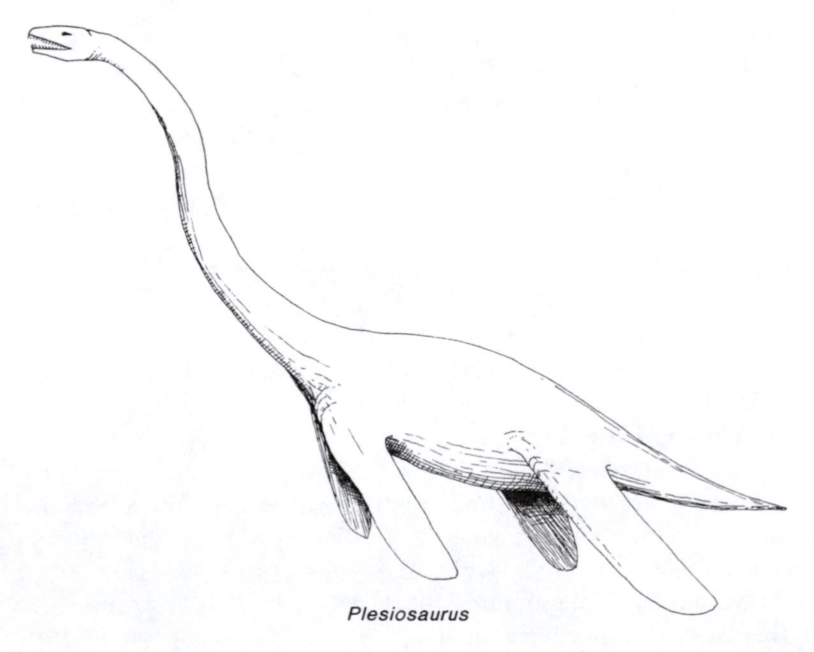

Plesiosaurus

Figure 11.13. Plesiosaurs such as this were abundant in Jurassic seas. These animals represent a group of reptiles that returned and became secondarily adapted to a marine environment. They ranged up to 10 metres long.

Flying reptiles, or pterosaurs, flew and soared in Jurassic skies. Lacking feathers and other typical bird characteristics, they were true reptiles that had adapted to conditions of flight. Their airfoil, or wing, was a thin membrane stretched between the body and the extremities of elongate limbs. Jurassic pterosaurs had wingspans of slightly over one metre.

JURASSIC TECTONICS

A major change in the pattern of plate motion takes place in the Jurassic as Pangaea fragmented. The westward-moving North American Plate formed a convergent margin with the eastward moving Pacific Plate. A subduction zone developed and produced the Franciscan mélange deposits. Associated massive igneous activity, which filled the eugeosyncline with volcanic debris and emplaced the granites associated with the Nevadian Orogeny.

Figure 11.14 illustrates the whole earth perspective of the break-up of the supercontinent. The widening Atlantic Ocean is shown, as is the connection between North America, Greenland, and Europe. A summary of plate motion is shown in cross-section on figure 11.15.

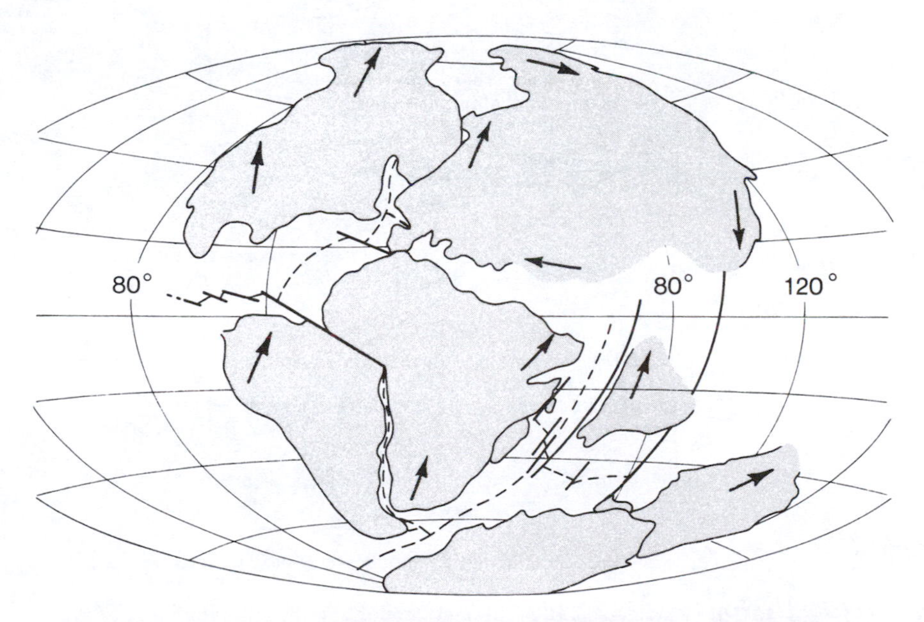

Figure 11.14. Diagram showing continental distribution during Late Jurassic. (After Dietz, R. S. and Holden, J. C., Scientific American, October 1970)

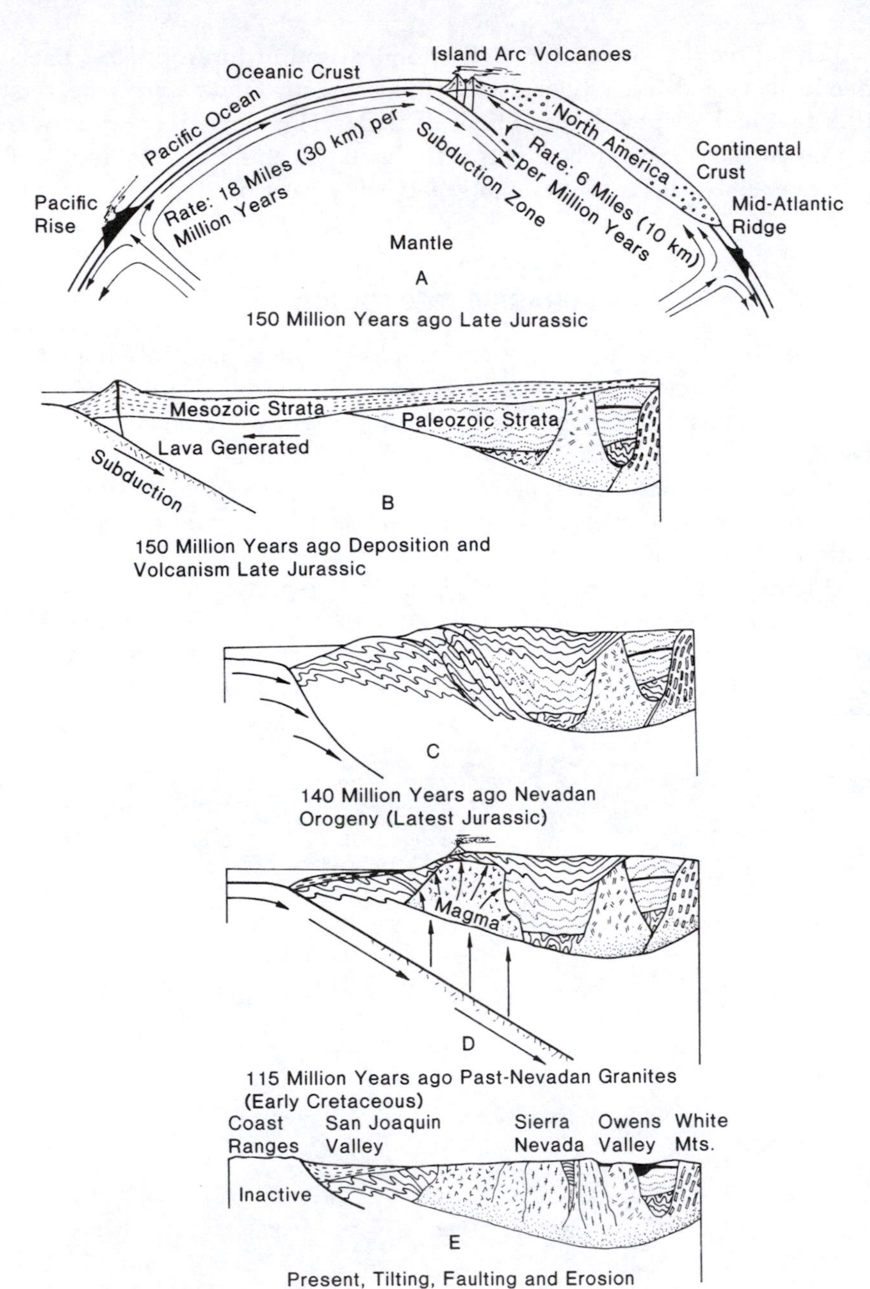

Figure 11.15. Cross section of plates throughout the development of the Nevadian Orogeny. (After Mintz, L. W., 1977, Historical Geology, 2nd Ed., C. E. Merrill, Publ.)

12

Cretaceous

TOPICS

The stratigraphic record of the Cretaceous is one of the most widespread of all geologic periods. Late Cretaceous marine deposits are more extensive than those of any other period since the Ordovician. The uplift of western America, which started in the Jurassic, continued throughout the Cretaceous, ultimately forming mountains from eastern California to Denver, Colorado. Continental uplift and igneous activity in western America reached a maximum in the Cretaceous. Volcanoes appeared for the first time in Cuba and the Antilles, and erupted in isolated spots in the stable interior of the continent from near Montreal, Canada, to Arkansas (fig. 12.1).

Sedimentation was widespread along the Atlantic and Gulf coasts, accumulating in thick aprons of material, which along with underlying Jurassic strata form the foundations of the modern coastal plains. Marine waters covered the western interior of the continent in Late Cretaceous time as seas transgressed southward from the Arctic and northward from the Gulf of Mexico to form a gigantic island sea, which flooded the areas of the present day Rockies and Great Plains. The Western Interior seaway is very similar in pattern, to the earlier Sundance-Zuni Sea, but was much larger in areal extent. During this widespread flooding of the continental interior, North America was divided east and west, into two large islands (fig. 12.2).

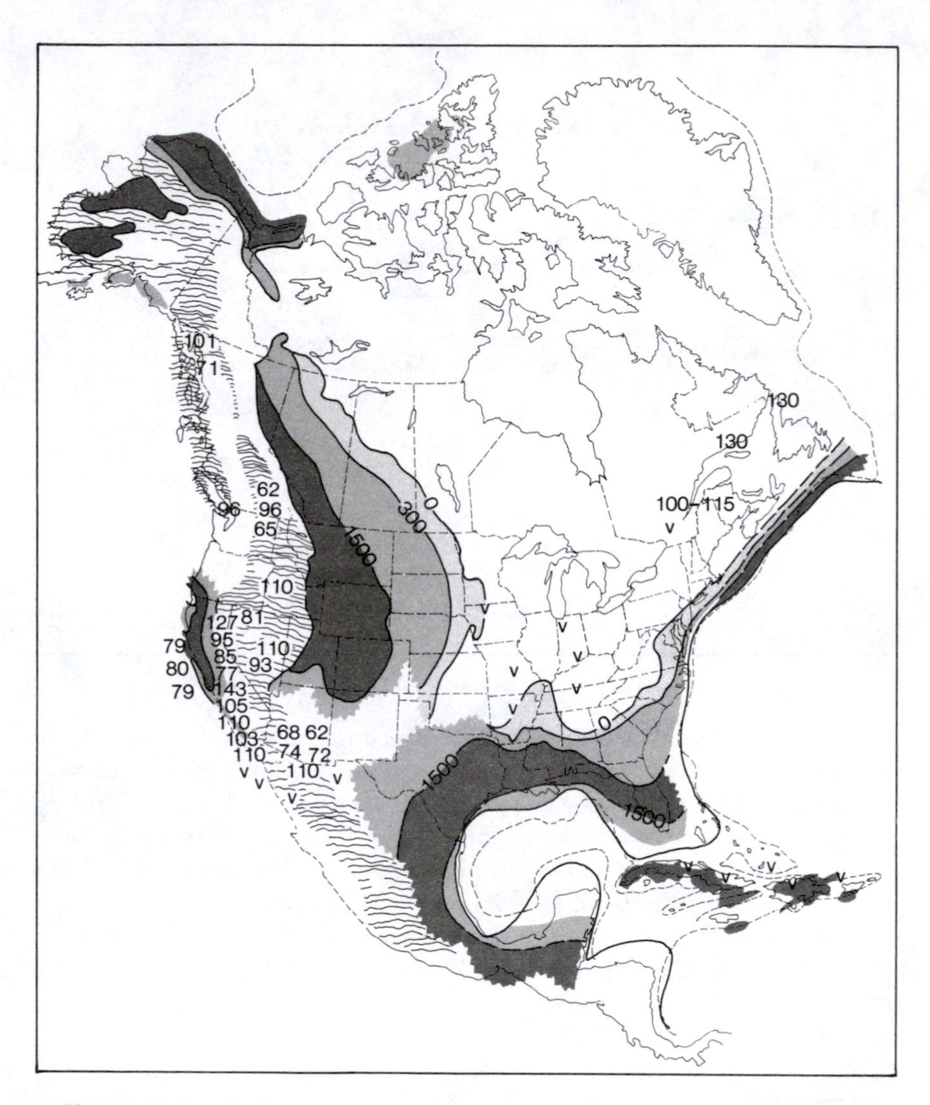

Figure 12.1. Thickness map of Cretaceous System in North America. Thickness in metres. Numbers scattered in western cordillera represent radiometric ages in millions of years. "V" represents volcanic rocks in the west and Antilles and cryptovolcanic structures in the central United States.

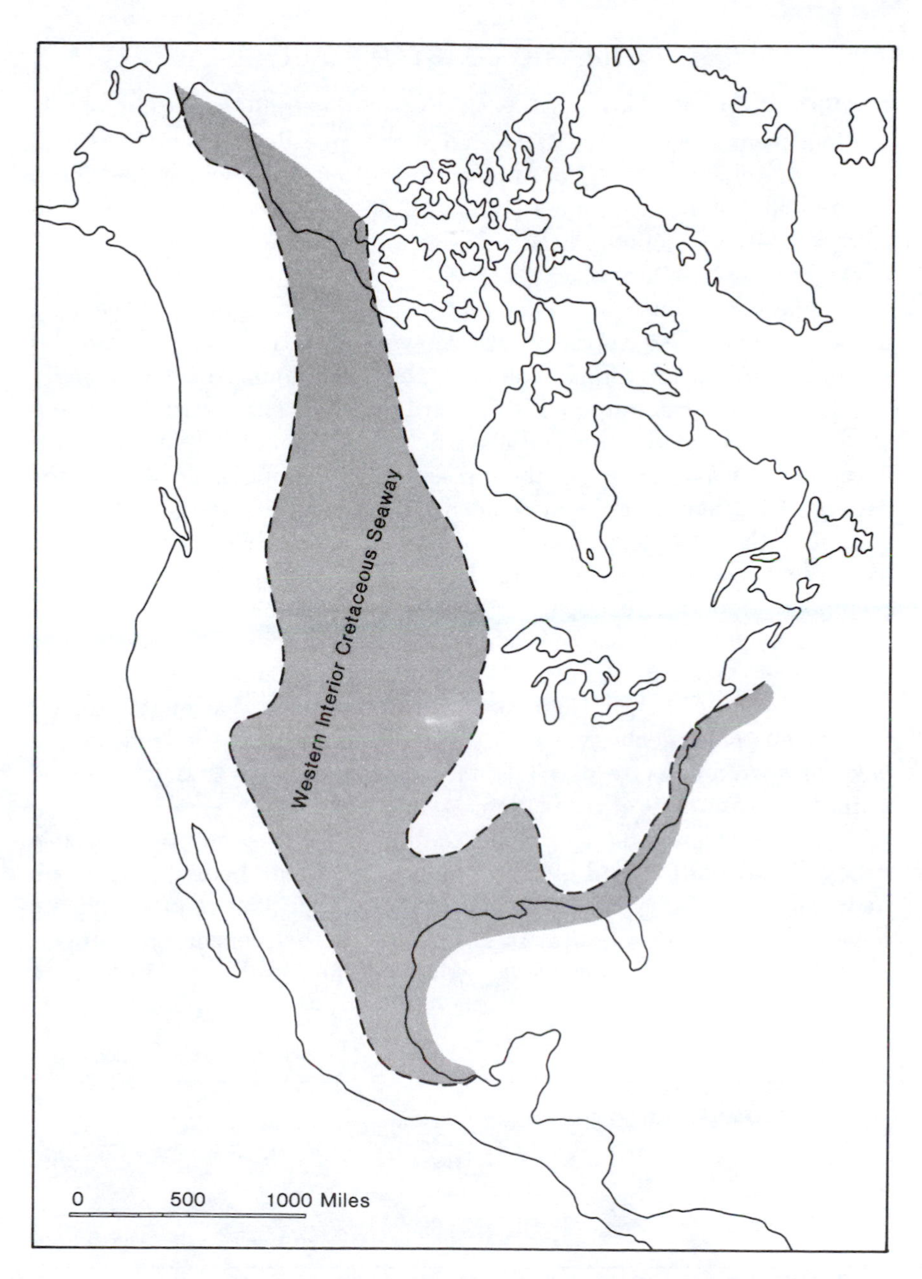

Figure 12.2. Marine seaway over North America during the Cretaceous. (After McGookey, D. P., 1972, p. 190. Geol. Atlas of Rocky Mtn. Region, Rocky Mtn. Assoc. Geol.)

ATLANTIC COASTAL PLAIN

With North America moving westward, the trailing margin of the continent, our present-day Atlantic coast, became flooded by the waters of the Atlantic Ocean. The Appalachian Mountains, which had been uplifted in the Pennsylvanian-Permian Allegheny Orogeny, were sediment sources for deposits accumulating along the Atlantic coast in the Cretaceous as they are today.

Cretaceous sediments are exposed in a belt extending from Georgia to Long Island, New York, and are known from marine dredge samples northeastward to the Grand Banks of Newfoundland. Along this entire belt Cretaceous strata dip gently seaward and thicken seward. They consist of stream-carried sands and clays deposited along a Cretaceous costline, which was similar to the present-day Atlantic coastline in the variety of its depositional environments. In the southeastern part of the continent, the Cretaceous coastline extended inland 160 kilometres from its present position, covering igneous and metamorphic rocks of the piedmont. Modern streams, which pass from resistant rocks of the piedmont on to the more easily eroded Cretaceous sands and clays of the coastal plain, increase in slope markedly along what is called the "fall line." Augusta, Georgia; Columbia, South Carolina; Washington, D.C.; and Philadelphia, Pennsylvania — all located on or near the "fall line" — have prospered because of availability of water power and from being at the upstream limit of navigation.

On the seaward side, Cretaceous sediments are covered by Cenozoic deposits, and both extend offshore. Since Cretaceous beds dip seaward, they soon extend beyond the reach of present drilling depths and are traced chiefly by geophysical means. While both Cenozoic and uppermost Cretaceous beds behave as relatively unconsolidated sediments,

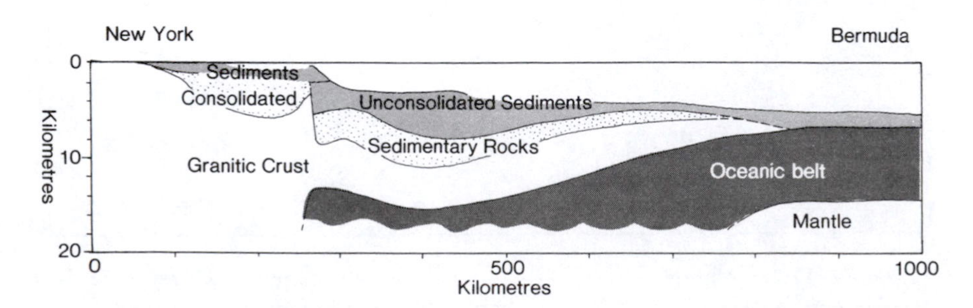

Figure 12.3. Cross section of the Atlantic Coastal Plain and adjacent rocks beneath the ocean floor from New York southeasterly to Bermuda, as determined from geophysical measurements.

most of the Cretaceous strata respond to seismic energy as semiconsolidated sedimentary rocks and can be followed offshore to considerable depths, as shown on figure 12.3.

An exploratory well, drilled to a depth of 5,000 metres on Sable Island, 300 kilometres east of Nova Scotia, penetrated more than 3,600 metres of Cretaceous strata and was still in Lower Cretaceous rocks at its bottom, indicating the substantial thickness of this apron of Cretaceous sediments on the continental shelf.

CRETACEOUS VOLCANISM IN EASTERN INTERIOR

Except for the coastal plain strip and two small patches of thin lake and stream deposits near Hudson Bay, Cretaceous sediments are absent from eastern North America. Near Montreal, Canada, the series of small intrusive bodies shown on figure 12.4 probably represents eroded remnants of volcanic necks, and yield radiometric dates between 100-115

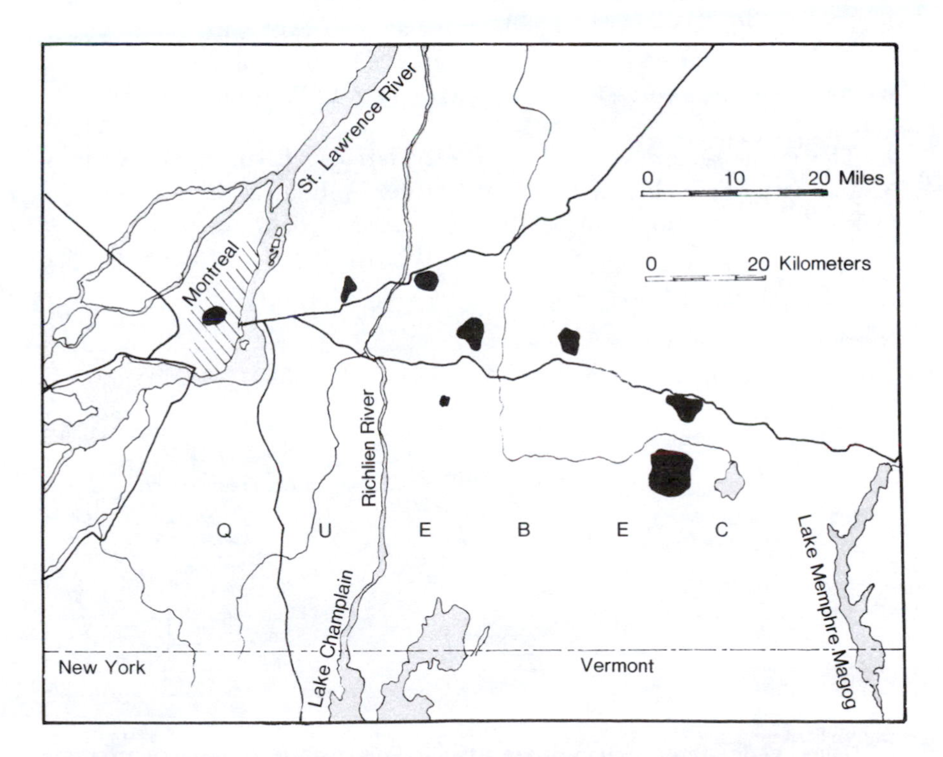

Figure 12.4. Cretaceous igneous rocks (in black) near Montreal, Canada.

million years, or Middle Cretaceous. Dikes of basic igneous rocks of Late Jurassic and earliest Cretaceous ages occur further eastward in maritime Canada, as indicated by the radiometric dates of figure 12.1. In east-central United States, small circular features called "cryptovolcanic" structures are scattered over several states. The name means "hidden volcanic" and implies that these structures represent ancient explosive volcanic activity; some of them have recently been interpreted as meteorite craters. Cretaceous age has been suggested for several of these features.

GULF OF MEXICO

Cretaceous sediments rim the Gulf of Mexico from Florida, where they occur in the subsurface to the Mexico-Guatemala border as shown on figure 12.1. They form an apron, similar to that along the Atlantic coastal plain, of fine-grained clastic sediments deposited in relatively shallow waters along the continental margin. Chemical and organic limestones, clay limestones, and chalks form a significant part of the Cretaceous deposits, particularly in Florida and Mexico. Figure 12.5 is a cross section illustrating Jurassic and Cretaceous strata thickening towards the Gulf.

The open eastern side of the nearly circular basin of the Gulf of Mexico is partially blocked by the Greater Antilles Islands: Cuba, Jamaica, Hispaniola, and Puerto Rico. Cretaceous rocks are prominent on all of these islands, and are somewhat different in origin than those that

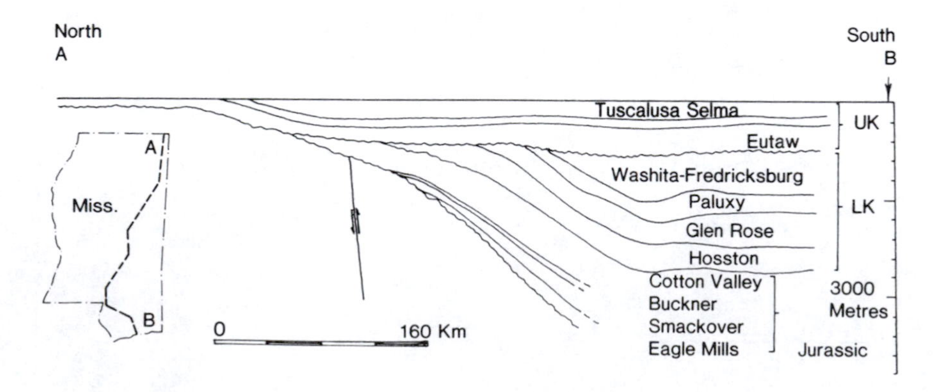

Figure 12.5. Cross section of Gulf Coastal Plain in Mississippi showing how Upper Cretaceous strata extend across both Jurassic and Lower Cretaceous units.

rim the other sides of the gulf, for they include substantial amounts of Cretaceous volcanic rocks, andesites, and basalts and graywackes. Along the northern edge of Cuba, limestones similar to those of Florida were deposited in Cretaceous time, suggesting a connecting shelf between the island and the continent.

Cretaceous rocks of the gulf area have yielded important oil production, particularly in Texas and Mexico. Interfingering sands and muds deposited in shallow water along the coast have provided a variety of stratigraphic situations suitable for the accumulation of oil. The oil itself is derived from microscopic plants and animals that flourished in warm, shallow waters and became a part of organic-rich mud deposits. Associated sands may be beach sands, channel sands, or offshore bars. After a pile of these various sands and muds has accumulated, the weight of the upper layers forces the oil out of the easily compressed muds into the pore spaces in the less-compressible sands, where it remains until reached by the drill. One of the largest such oil fields in the United States is shown in cross section (fig. 12.6).

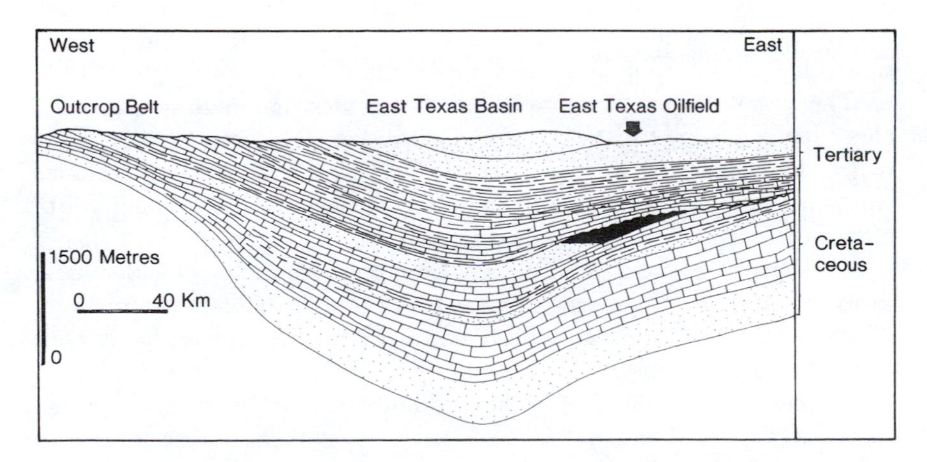

Figure 12.6. Cross section of East Texas Oil Field showing oil trapped in Cretaceous sandstones unconformably below uppermost Cretaceous and Tertiary rocks. (After Halbouty, M. T., Am. Assoc. Petrol. Geol. bull. v. 56, p. 537-541, 1972)

CRETACEOUS MOUNTAIN BUILDING

Two basic patterns of deformation are recorded in Cretaceous rocks of western America. The first is compressional, expressed in the rocks as intense overthrust faulting and folding, which extended from Late Jurassic through the Cretaceous. This phase of tectonism is called the Sevier Orogeny in the United States and Columbian Orogeny in Canada. A second wave of mountain building followed the Sevier Orogeny and is named the Laramide Orogeny after the Laramie Mountains of Wyoming, where the effects of this Late Cretaceous to Early Tertiary tectonic event are well displayed. The Laramide Orogeny is characterized by a mountain-building pattern of scattered vertical, sometimes asymmetric, uplifts located generally east of the Sevier Mountains. These mainly include the middle and southern Rockies of Colorado and Wyoming, east to the Black Hills in South Dakota. As a name for the combined Sevier and Laramide Mountains, Cordilleran Orogeny is used by many authors. A summary of these events is illustrated in figure 12.7. The Sevier Mountains are illustrated as highlands on figure 12.1, where they extend from Alaska to Mexico. The Laramide Mountains are shown on figure 12.8. A cross section of the Uinta Mountains of eastern Utah is illustrated in figure 12.9 as an example of Laramide uplifts.

The compression that produced the Sevier mountains originated along the convergent plate boundary at the Pacific coast and moved the rocks, shinglelike, toward the east. In some places the original distance was shortened as much as 160 kilometres, shortening the original width of these crustal rocks by 50 percent in some cases.

Throughout western America are numerous examples of older rock, thrust upon Late Cretaceous strata. For example, Chief Mountain in Glacier-Waterton Park was moved eastward by the Lewis thrust fault approximately 75 kilometres, placing Precambrian rock on top of Late Cretaceous shales. Considering the age and structural relations now existing at Chief Mountain, as well as many other similar situations throughout the Rocky Mountains, one can better understand the Late Cretaceous timing of the Sevier Orogeny. Figure 12.10 illustrates the narrow, continuous belt of overthrusting in Western America. This western overthrust belt is currently the site of intense petroleum exploration and extensive production, as a result of recent major discoveries in east central Utah and southwestern Wyoming.

We use three separate names to identify Mesozoic orogenic events in western America: Nevadian for the western part. Sevier for the middle part, and Laramide for the eastern belt. The age distribution for these

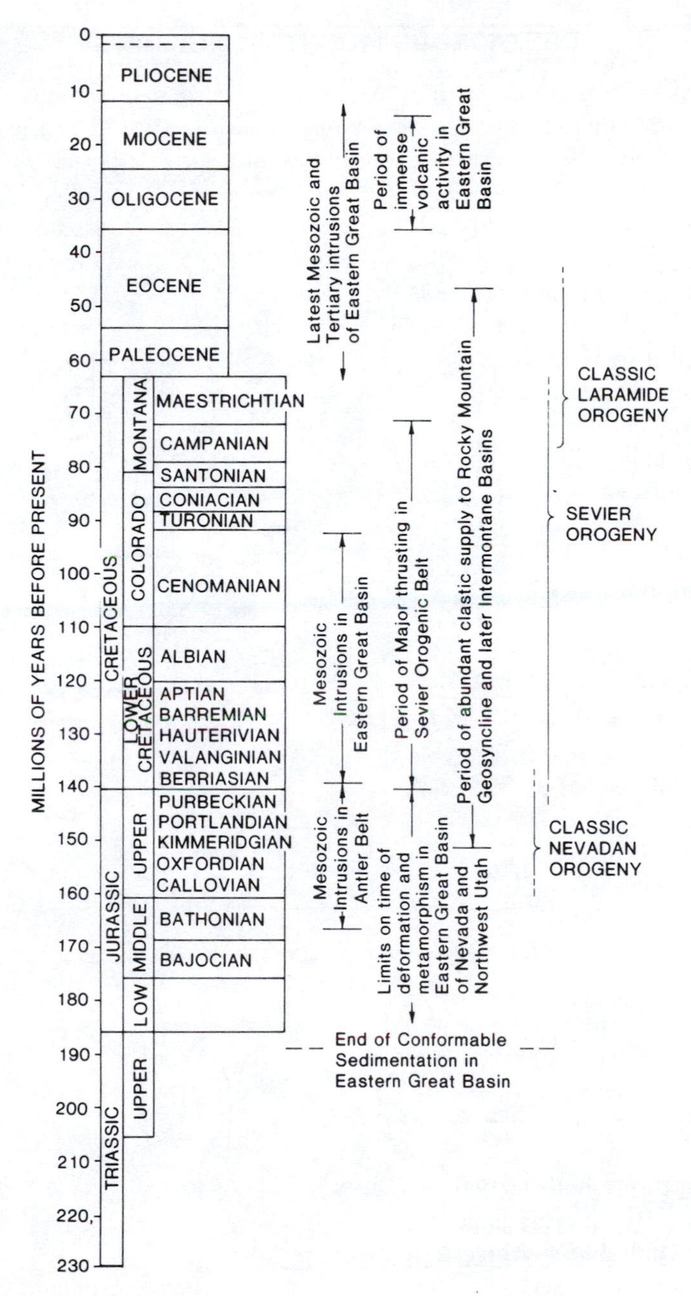

Figure 12.7. Chart showing age relationships of Mesozoic orogenies. (After Armstrong, R. L., 1968, Geol. Soc. Amer. Bull. 79:452)

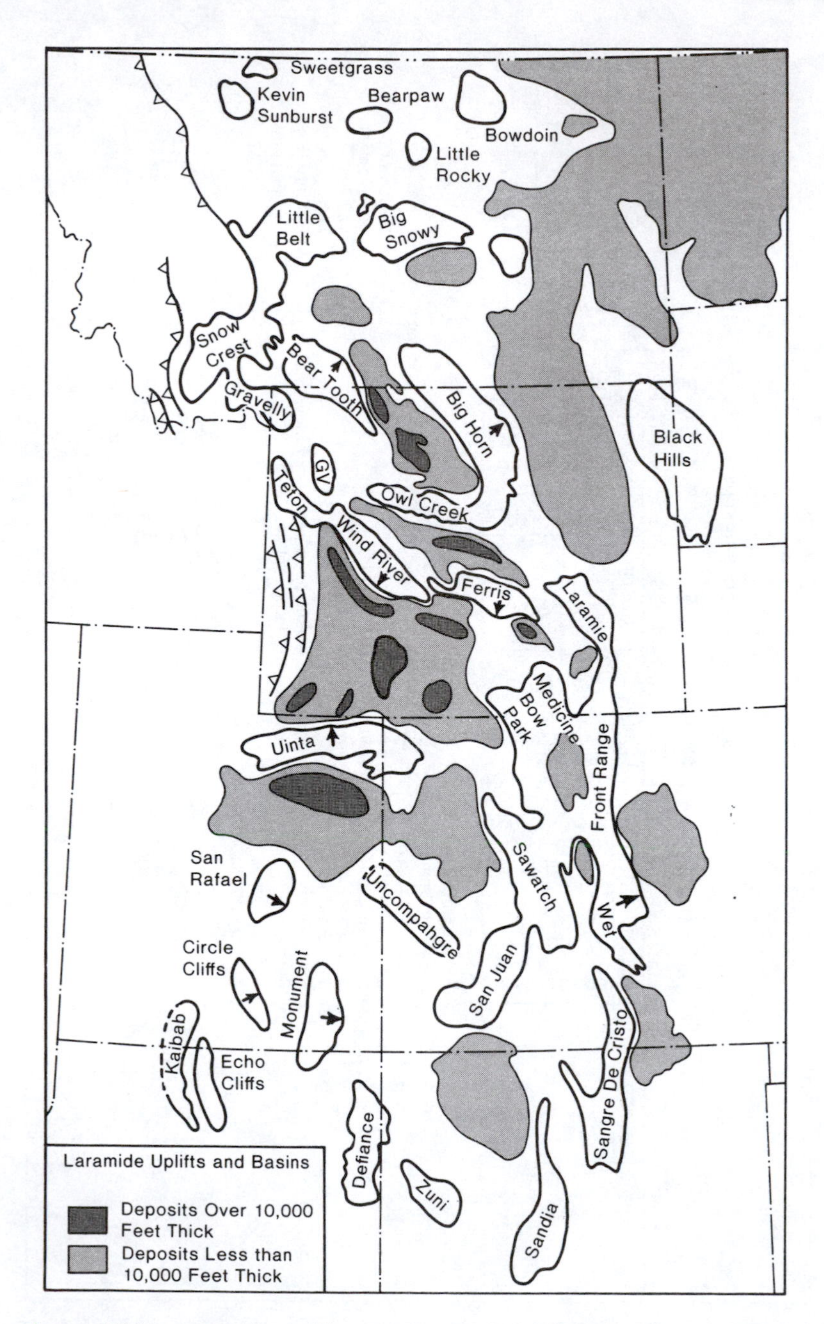

Figure 12.8. Laramide uplifts in western United States. Arrows point in the direction that each upfold has moved, as indicated by its asymmetry or direction of overthrust.

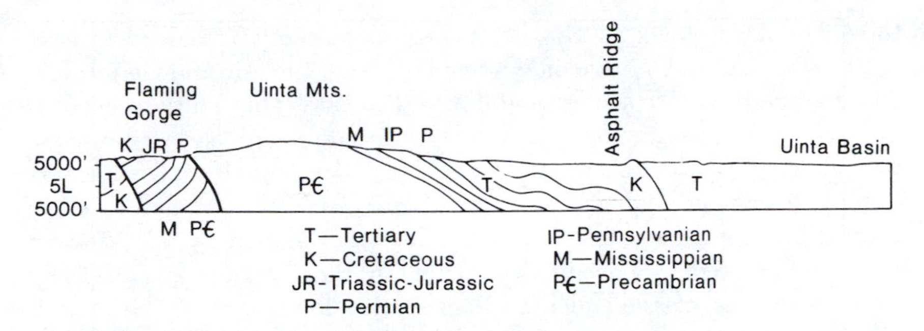

Figure 12.9. Cross section showing structures formed during Laramide Orogeny on east side of Uinta Mountains, Utah.

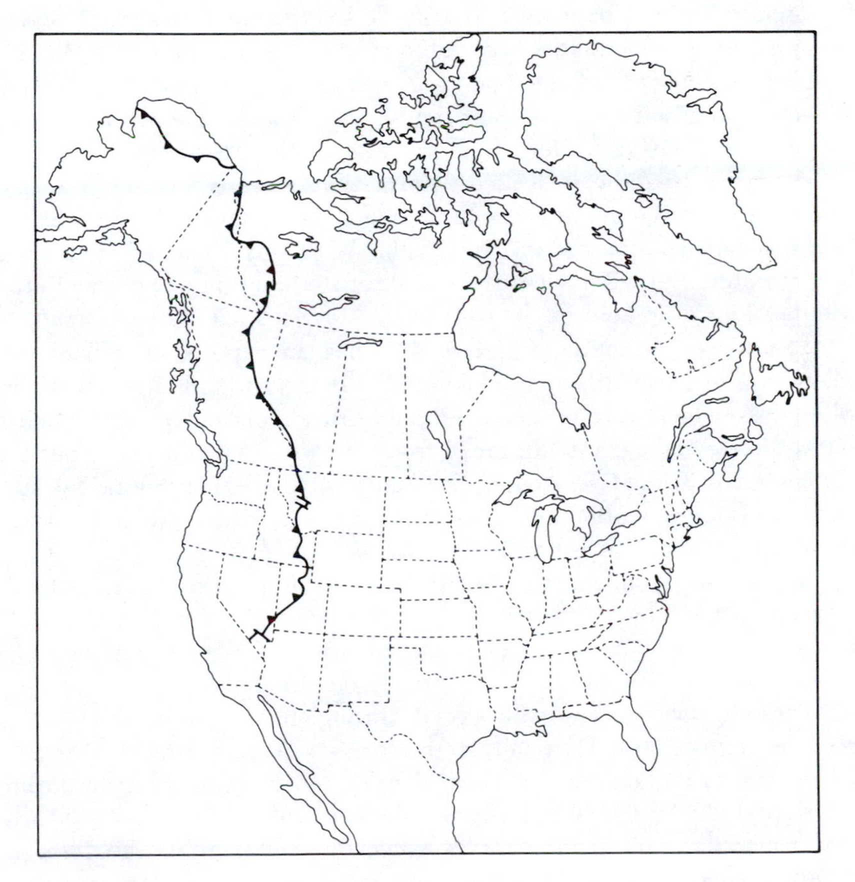

Figure 12.10. Map showing overthrust belt in Western United States and Canada.

three Mesozoic orogenies also fits a general pattern; from west to east, the Jurassic Nevadian is the oldest, Sevier is Late Jurassic and Cretaceous, and Laramide is Cretaceous and Early Tertiary. This pattern suggests a more or less continuous wave of deformation, which progressively moved eastward through time. As we shall discuss in the following section of this chapter, when the Sevier Mountains were well developed in central Utah, eastern Utah and Colorado were flooded by the Western Interior Cretaceous Seaway. Only later in the Cretaceous and Early Tertiary did the classic Colorado Rockies develop.

CRETACEOUS BATHOLITHS AND VOLCANICS

Cretaceous igneous activity is recorded throughout western America by large granitic batholiths and widespread volcanic debris (fig. 12.11).

Large batholiths of Cretaceous age are present in Idaho and Montana, as well as in granitic intrusions of the Sierra Nevadas. The main pulse of granitic intrusion in western America was in the Mid-Late Cretaceous (80-90 m.y.). Like the associated deformation, igneous intrusion shows a general eastward migration through time. Of economic interest is the fact that much of the metallic ore in the West was "steamed" into their host rock during the Cretaceous Period, for example the copper at Butte, Montana, and the gold of the California Mother Lode.

The Idaho Batholith (figure 12.11) has an exposed length of 400 kilometres and a width of 130 kilometres. It is surrounded by metamorphic rocks, which were produced by the intrusion of magma. Radiometric dates on the Idaho Batholith range from 217 to 38 million years, but are concentrated in the Cretaceous. The Southern California, Sierra Nevada, and Crest Range batholiths of Washington and British Columbia were emplaced over the same time span.

In western Montana, the Boulder Batholith (figure 12.11) extends over an area 100 by 50 kilometres. Like the others, it is composed mainly of granitic rock capped by an ejecta blanket of its own making. The Boulder Batholith is similar in age to the one in Idaho.

Volcanic sources are widespread throughout western America, as seen on figure 12.11. Especially active centers were located in Nevada, Idaho, western Utah, and Montana. Eruptions were also common during later parts of the Cretaceous from volcanoes in Arizona, New Mexico, and Colorado. Volcanic tephra is especially common in the form of bentonite beds within the enclosing Cretaceous shales or sandstones. Bentonite beds, because they represent virtually instantaneous geologic

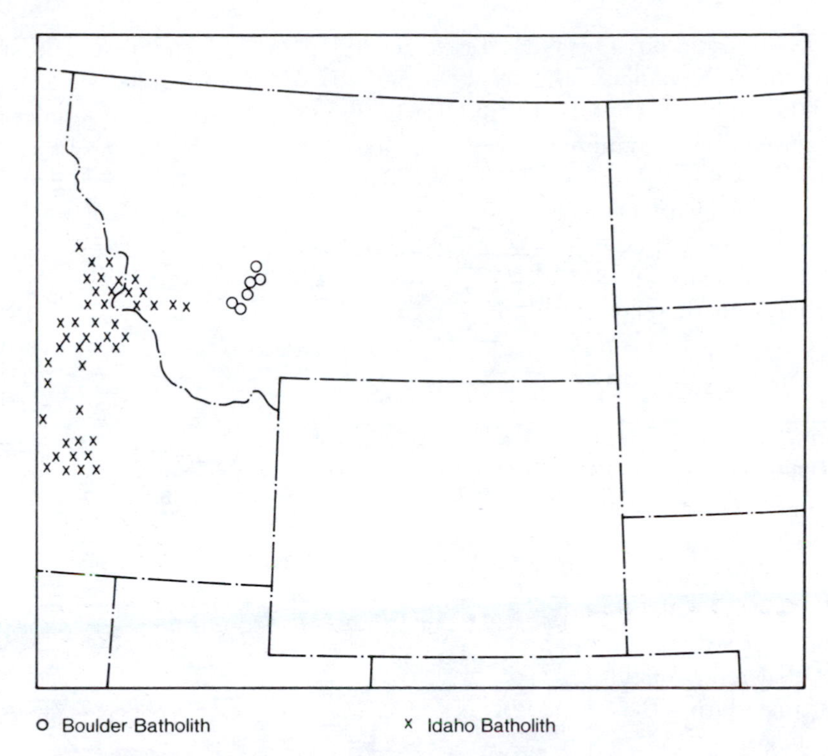

Fig. 12.11. Map of Cretaceous batholiths in Idaho and Montana. (After McGookey, D. P., 1972, Geol. Atlas of Rocky Mtn. Region, p. 192, Rocky Mtn. Assoc. Geol.)

events covering broad areas of the landscape, make excellent time markers for stratigraphic correlation. The bentonite has economic value in drilling mud, fillers, and as ingredient in modern steel-making processes.

CRETACEOUS SEDIMENTARY ROCKS OF THE WESTERN INTERIOR

Cretaceous sedimentary patterns in the west are dominated by contemporary mountain building, the interior seaway, or both. In many cases the dominance of one of these over the other shifts back and forth through time, leaving an interesting stratigraphic record. Figure 12.12 illustrates Cretaceous rocks deposited near the shore of the interior seaway and along the eastern flank of the Sevier Mountains, where non-

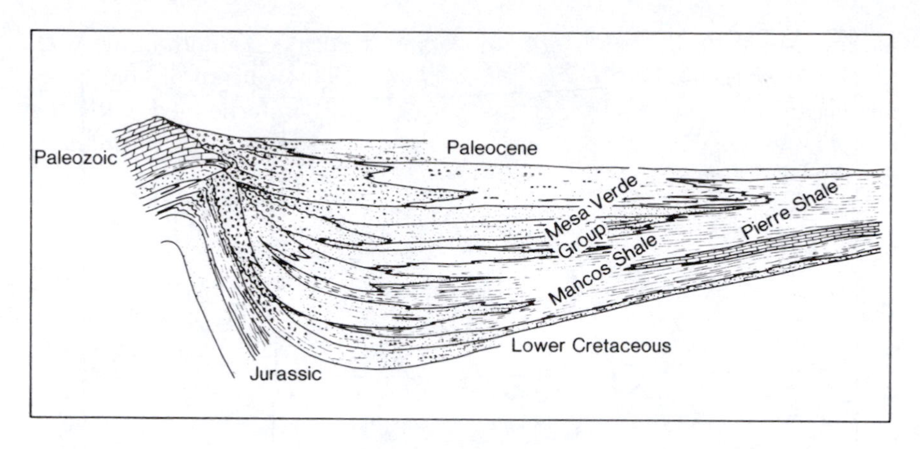

Figure 12.12. Cross section of Jurassic Cretaceous and Paleocene strata in Central Utah. (After Armstrong, R. L., 1968, Geol. Soc. Amer. Bull., 79:429-458)

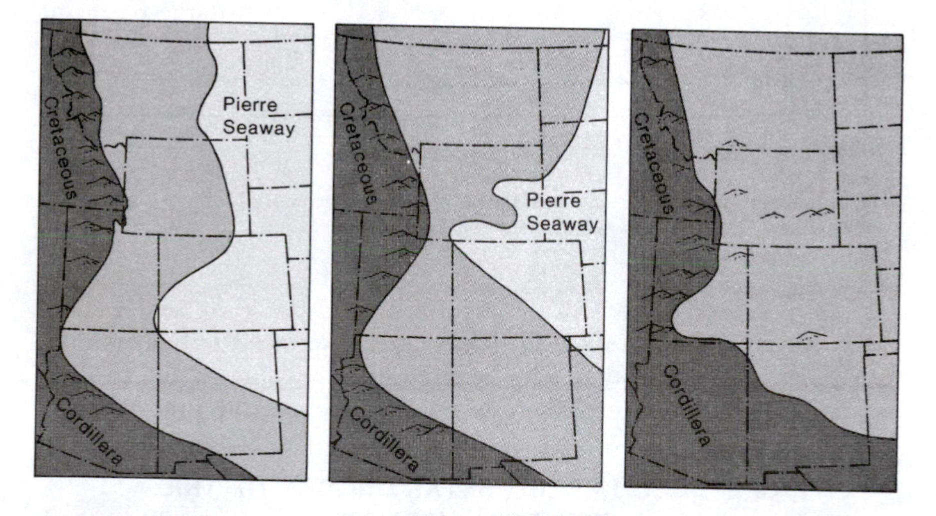

Figure 12.13. Maps showing development of Cretaceous marine and non-marine strata in Western America. (After McGookey, D. P., 1972, Geol. Atlas of Rocky Mtn. Region, p. 222, 225, and 227, Rocky Mtn. Assoc. Geol.)

marine Mesa Verde Sandstone shale and coal beds intertongue with fossiliferous marine beds of the Mancos Shale. This pattern of coal bearing clastic deposits shed from the Sevier Uplift, interbedded with the contemporaneous marine beds characterizes Cretaceous rocks throughout the Rockies (figure 12.13).

As a major economic deposit of western America, coal occurs in thick widespread units over much of the West in both the United States and Canada. On the Crow Indian Reservation in Montana, for example, coal occurs in three virtually continuous beds which total over 50 metres thick. These beds contain over 17 billion tons of high quality coal.

PACIFIC COAST

In northern California, southwest Oregon, and Vancouver Island, unmetamorphosed Early to Late Cretaceous shallow water marine sandstones, conglomerates, and dark shales indicate deposition along the Cretaceous west coast side of the Cordilleran orogenic belt. Large basins of Cretaceous shallow water marine and continental deposits are present in British Columbia and Alaska, as shown on figure 12.1.

ARCTIC COAST

Cretaceous sediments in northern Alaska and adjacent Canada form a coastal plain apron that thickens to 7,000 metres northward toward the Arctic Ocean. Relations are nearly identical with those on the Atlantic coast in that sediments are mostly mudstones, sandstones, and pebble conglomerates, which were derived from rising lands just to the south of the coastal plain.

CRETACEOUS LIFE

Ammonoids (fig. 12.14) again dominate Cretaceous marine invertebrates. Before their extinction late in the period, they evolved into numerous types including some bizarre forms of most unusual character. These so-called heteromorphs represent the final ammonoid occurrences ending a history of unexcelled usefulness as time indicators, which started in the Devonian. The absence of ammonoids in Cenozoic strata assists geologists in distinguishing Mesozoic and Cenozoic deposits.

Bivalves (fig. 12.15) are common in Cretaceous deposits as are belemnoids and echinoderms. Corals and bryozoans, also abundant during the period, resembled their modern counterparts.

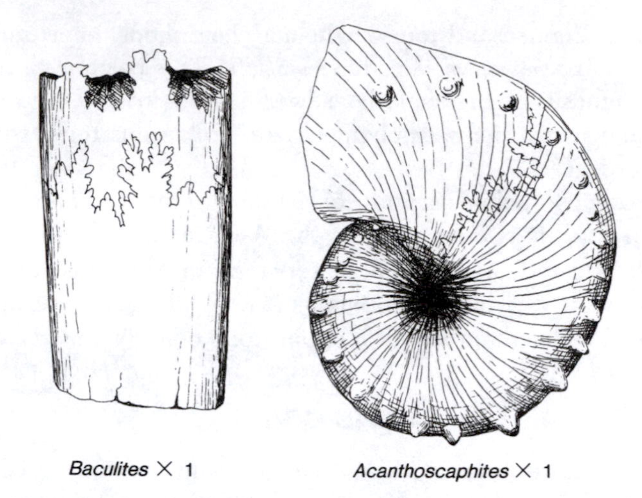

Figure 12.14. Ammonoid cephalopods showing complex suture pattern typical of Cretaceous forms. **Baculities** was a long, thin ammonoid, coiled only at one end. Only part of **Baculities** is shown here since this is the way they occur in most collections. **Acanthoscaphites** is typical of most coiled Cretaceous forms.

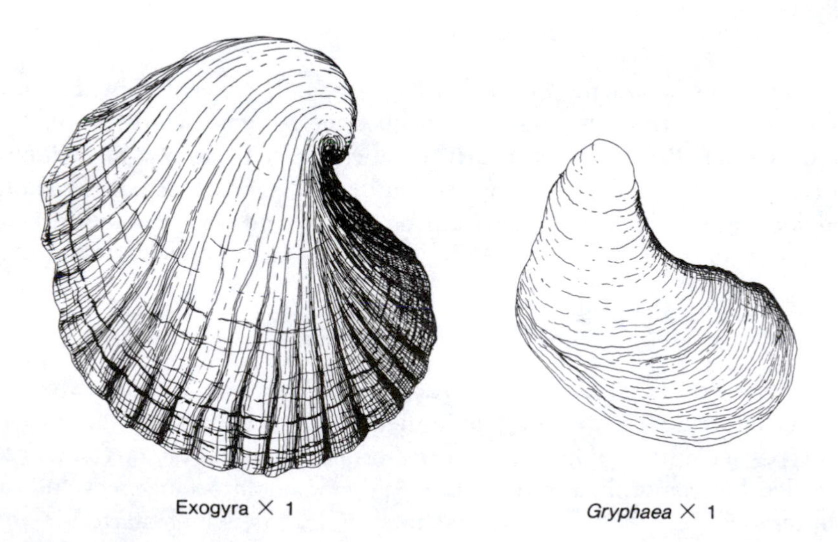

Figure 12.15. Cretaceous bivalves. Remains of these thick-shelled oysters are abundant in most marine Cretaceous strata.

Dinosaurs reached the climax of their domination in a giant carnivore named *Tyrannosaurus* (fig. 12.16) nearly seven metres in height, 15 metres long, and weighing up to 10^4 kilograms. The spiked tail and plated back of *Stegosaurus* seems an obvious defense modification against such voracious predators. The most successful dinosaur in terms of survival seems to have been rhinoceroslike *Triceratops* (fig. 12.17) whose fossil remains are found in the youngest of all dinosaur-bearing strata.

Ichthyosaurs and plesiosaurs, and a 15-metre-long crocodilelike mosasaur continued to dominate marine life. Ammonoids were apparently acceptable food for mosasaurs, as indicated by a tooth punctured shell found in a Cretaceous marine deposit. Pterosaurs, the flying reptiles of the Mesozoic, attained wing spans up to 18 metres, based upon a recent discovery of an unusually large specimen discovered in south Texas. Cretaceous chalk beds of Kansas also have yielded many specimens of the airborne reptiles. Marine turtles of the Cretaceous measured up to four metres in length. Cretaceous mammals, although developing toward modern types, remained small amid their large reptilian predators.

Cretaceous plant evolution is characterized by the explosion development of angiosperms, or flowering plants. Although rare in Lower Cretaceous strata, their fossils are dominant in the Late Cretaceous. Estimates that as much as 90 percent of Late Cretaceous plant fossils are

Figure 12.16. Tyrannosaurus, of Cretaceous age, is probably the most fearsome land carnivore that ever lived. This reptile stood more than 6 metres high. Its forelimbs were so small as to be nearly useless.

Triceratops

Figure 12.17. Triceratops, a Cretaceous herbivore, had the distinction of carrying the most massive head known in land animals. The protective shield on the largest specimens measures 2 metres across.

angiosperms. Representatives of this group, which appeared and attained essentially their modern form in the Cretaceous time, include the first grasses, and the following trees: fig, sassafras, magnolia, birch, walnut, poplar, holly, ebony, elm, and oak. These plants provide the food for many groups of animals, such as bees, wasps, and butterflies, which came into existence as a result of the availability of new plant foods.

CRETACEOUS TECTONICS

Throughout Cretaceous time, North America continued to move westward, as shown in figure 12.18. A subduction zone existed along the leading or convergent margin of the continent and provided the compression for the Sevier tectonism, and the magma sources for Cretaceous igneous activity (figure 12.19). Figure 12.20 shows the expansion of the Atlantic Ocean basin and the incipient separation of Greenland from North America. This general pattern of the break-up of the supercontinent continues to date.

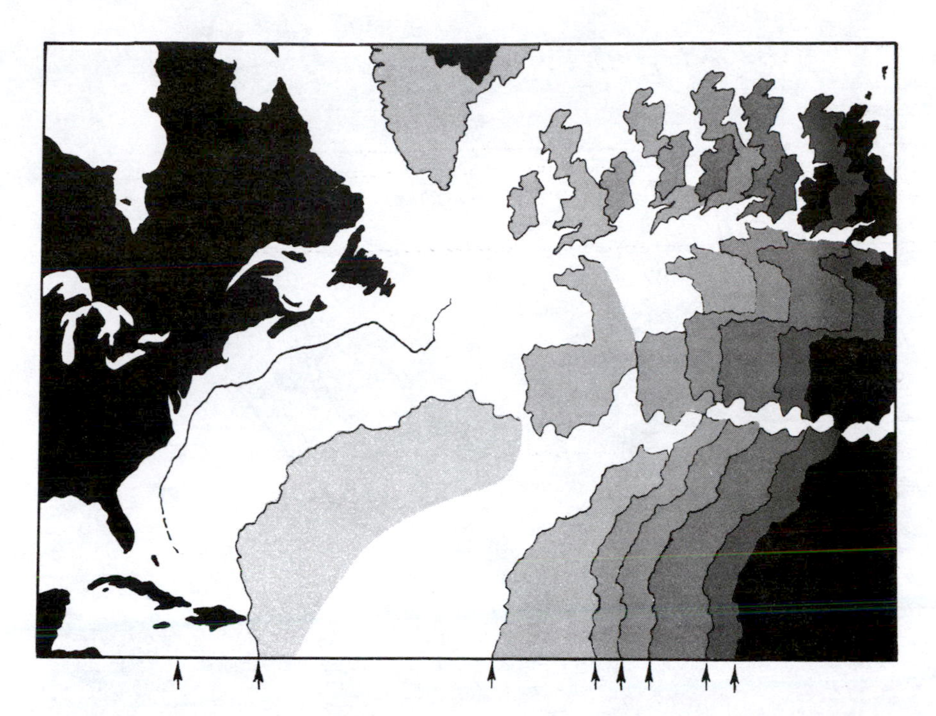

Figure 12.18. Relative positions of Europe and Africa with respect to North America from Triassic time until the present. (After Pitman, W. C. and Talwani M., Geo. Soc. Amer. bull. v. 83, p. 629, 1972)

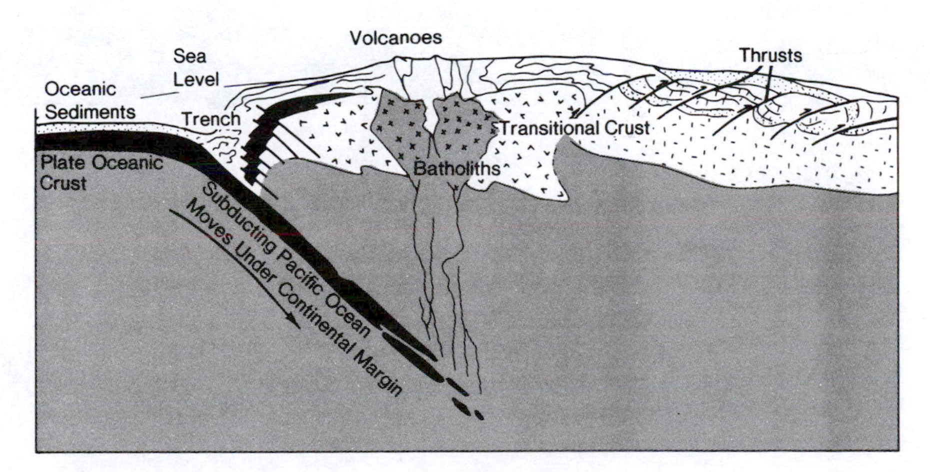

Figure 12.19. Cross section of plate relationships along Western North America during the Cretaceous Period. (After Dewey, J. F. and Bird, J. M., 1970, Jour. Geophy. Res., 75:2638)

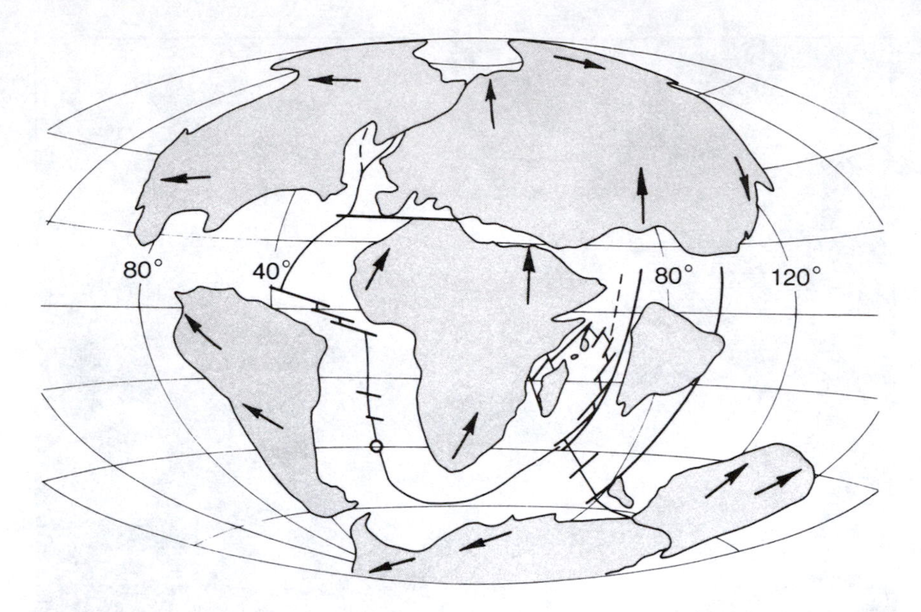

Figure 12.20. Diagram showing continental distribution during Late Cretaceous. (After Dietz, R. S. and Holden, J. C., Scientific American, October 1970)

13

Tertiary

TOPICS

The topography of our modern world came into being with the last 65 million years of geologic time as a result of major Tertiary modifications of older earth structures. Mountains, continental outlines, and most of the major drainage patterns, reflect Tertiary events. The Quaternary Period has accentuated Tertiary patterns of erosion and deposition by adding the distinctive mark of glacial processes to the Tertiary landscape.

CENOZOIC TIME SUBDIVISIONS

Events of the Cenozoic are so well preserved that we are able to more precisely subdivide this era than either the Paleozoic or Mesozoic. The Cenozoic is divided into two very unequal periods, the Tertiary and Quaternary. Of the 65 million years of Cenozoic time, the Quaternary includes about the last 2 million. Thus, the Tertiary Period embraces nearly all of the Cenozoic, and the terms "Cenozoic" and "Tertiary" are nearly synonymous. Cenozoic periods are subdivided into epochs (fig. 13.1), whose relative length has been determined by radiometric dating of both continental ash beds and submarine ash layers interbedded with fossil-bearing muds in deep-sea cores. The radiometric dating, as shown

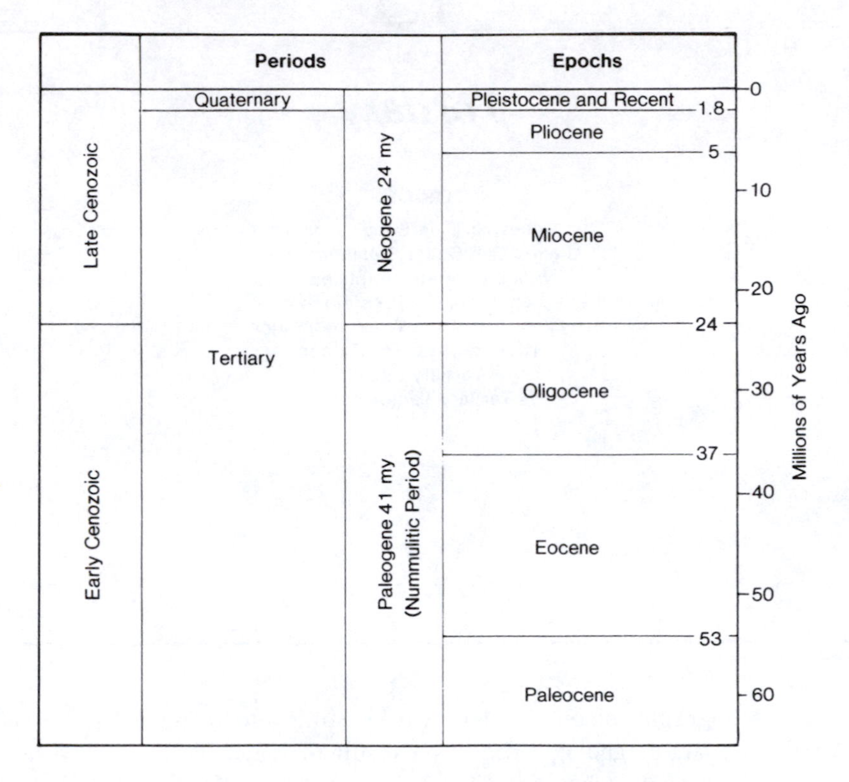

Figure 13.1. Cenozoic time subdivisions.

on figure 13.1, has made it possible to compare the age relationships of such widely separated groups as plains-dwelling mammals and marine foraminifera. The numbers obtained by radiometric methods are interpreted differently by various authors, consequently different dates are sometimes cited for Cenozoic epochs.

ATLANTIC-GULF COAST SEDIMENTATION

Figures 13.2 and 13.3 show the thickness of the apron of sand, silt, and mud derived from the continent and deposited on the eastern and southern continental shelves. As shown in figure 13.4, the deposits built continuously seaward as the water retreated.

Oil and gas are the most important economic products of Cenozoic strata in the Gulf Coast, although little has been found along the Atlantic

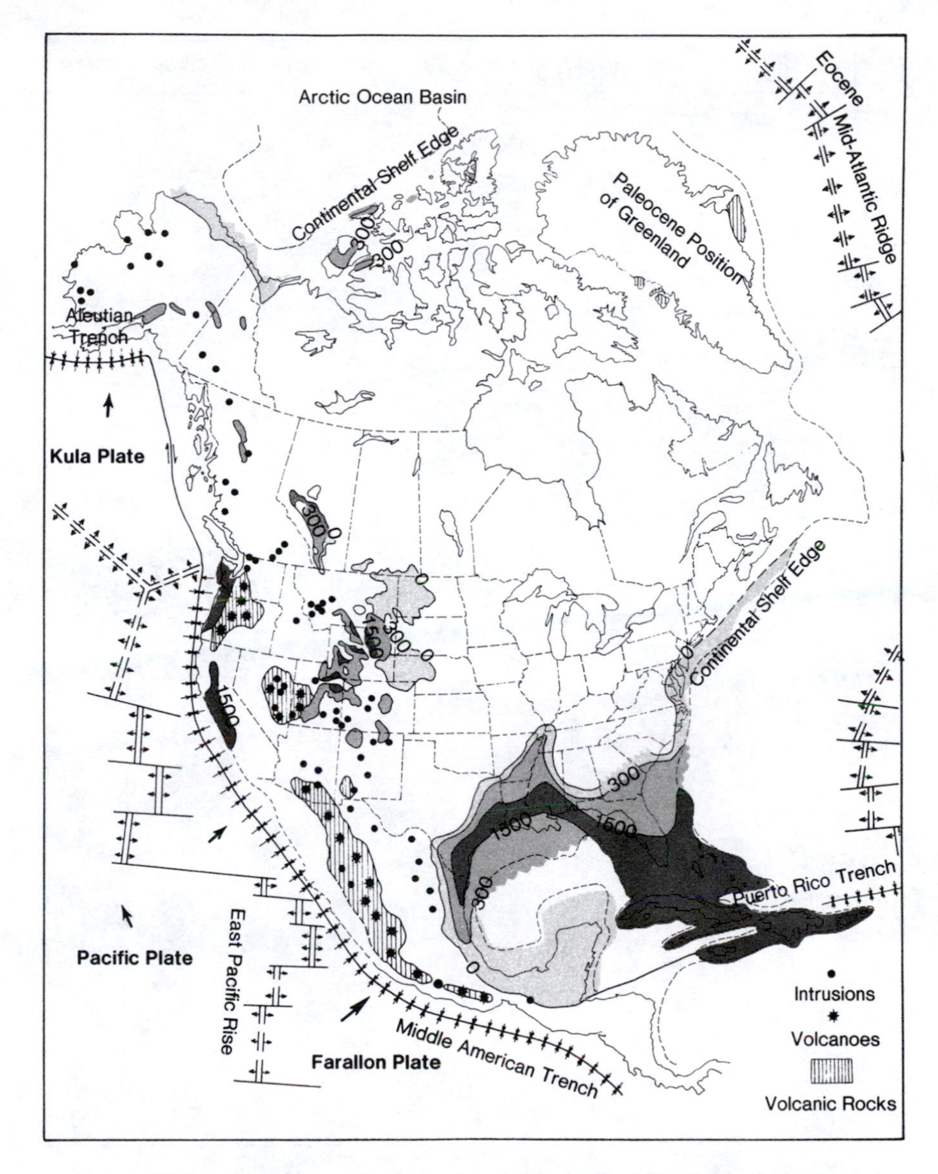

Figure 13.2. Early Cenozoic features of North America. Rock thickness in metres. Inferred plate relations. (After Atwater, T., Geol. Soc. Am. bull. v. 81, p. 3513-3536. 1970)

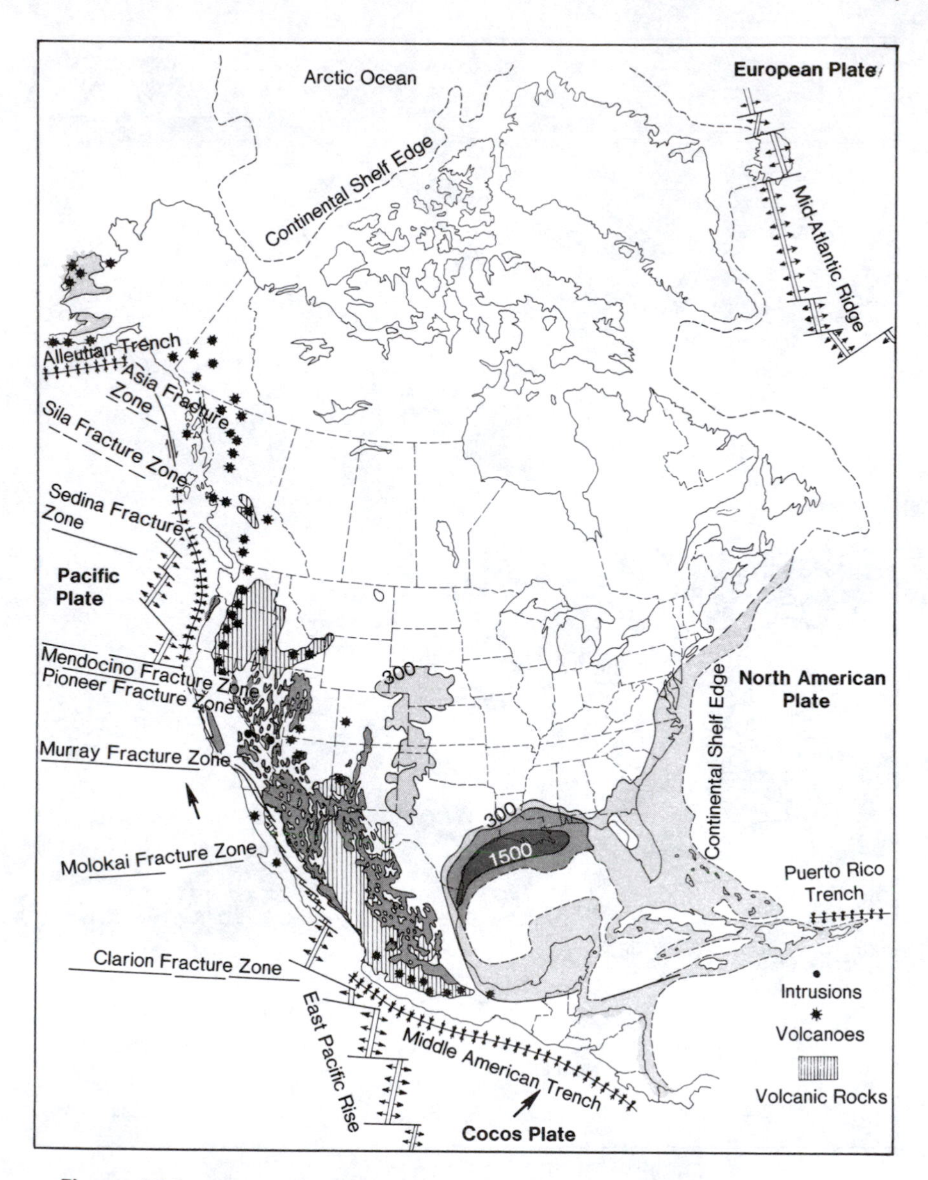

Figure 13.3. Late Cenozoic features of North America. Rock thickness in meters. Arrows show movement of oceanic plates relative to North American plate.

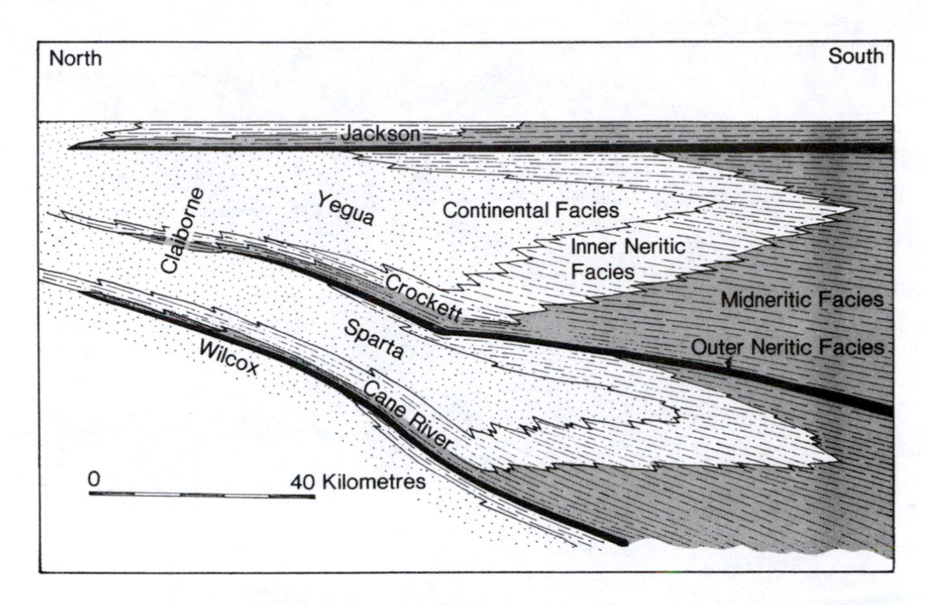

Figure 13.4. Interfingering of sedimentary rock types in the Tertiary of the Texas Gulf Coast. (After Lowman, S. W., Am. Assoc. Petrol. Geol. bull. v. **83**, p. 1939-1977, 1949) Continental strata grade from north to south into marine units in intertonguing relationships.

Coast. Hydrocarbons result from the interfingering of organic-rich muds and porous sandstones deposited along ancient coastlines. One of the most interesting types of petroleum reservoirs and traps is that related to salt domes (figs. 13.5 and 13.6), which are more numerous along the Gulf Coast than anywhere else in the world. The salt was deposited as a blanket in Jurassic time and buried by successive layers of sediment. The weight of the overlying sediment has squeezed the salt upward into Cenozoic sediments, because salt is lightweight and tends to "float" upwards in sediments. Oil accumulations of astonishing quantity have formed around many of the Gulf Coast salt domes.

FLORIDA-BAHAMAS-ANTILLES

Southern Florida and adjacent islands to the south and east are comprised chiefly of carbonate rocks. These are chemically and organically derived sediments that accumulated to great thickness in this area during the Cretaceous and Early Tertiary, as shown in figure 13.7. The thinness of the Late Tertiary deposits compared to the Early Tertiary in this area, as shown on figure 13.2 and 13.4, may be related to Quaternary changes

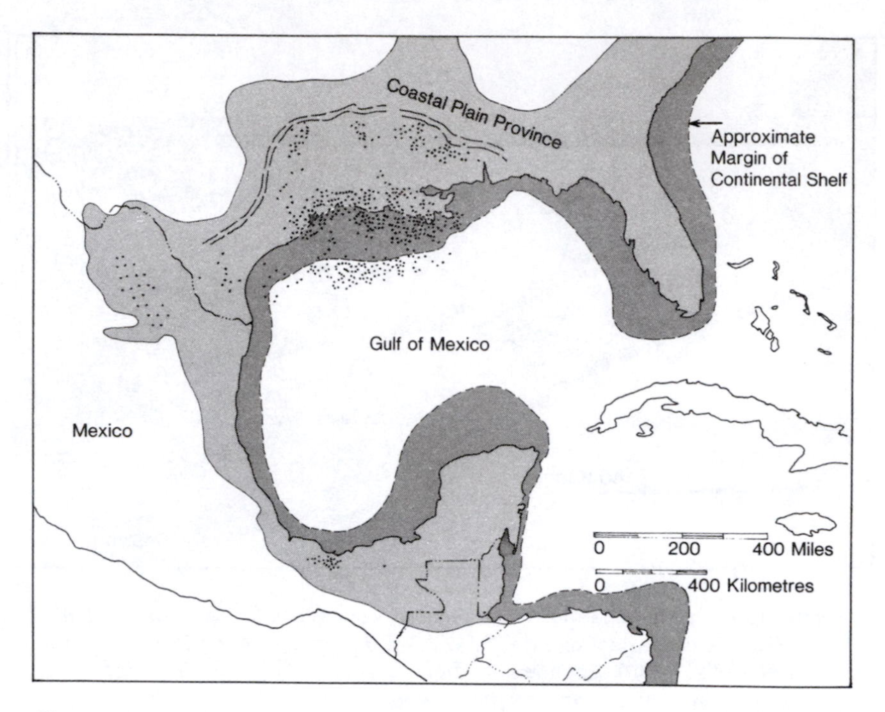

Figure 13.5. Distribution of salt domes along the Gulf Coast. Salt has risen from Jurassic layers upwards through Cretaceous and Tertiary strata.

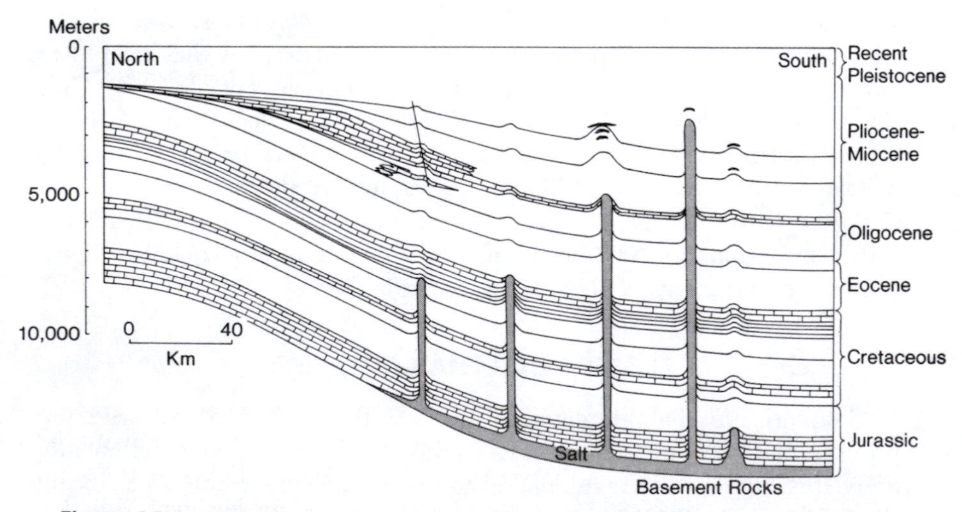

Figure 13.6. Salt domes showing deformation of beds adjacent to the rising column of salt. Salt domes in the Gulf Coast are about 2000 metres across. (After Carsey, J. B., Am. Assoc. Petrol. Geol.)

Figure 13.7. Bahama Islands, built of organically deposited limestones which accumulated to a thickness of many thousands of metres in Cretaceous and Tertiary time. This is a new portion of the continent built largely by organic growth and chemical precipitation. Numbers represent fathoms of water.

in sea level. During the Quaternary Period, sea level has been alternately one hundred metres higher or lower than at present. Lowering of the sea level by this amount exposed vast tracts of the Atlantic continental shelf to erosion, and any Tertiary sediments would have easily been carried far out toward the edge of the continental shelf, hence no longer available for inspection.

Florida, the Bahamas, and the Antilles, represent major new growth, by organo-chemical means, of part of North America. This carbonate deposit now forms a barrier that separates the Sigsbee Deep in the Gulf of Mexico from the remainder of the Atlantic Ocean basin of which it was once a part.

EARLY TERTIARY OF THE WESTERN INTERIOR

Laramide uplift of Rock Mountain ranges continued from late Cretaceous into earliest Tertiary time, indicated by folded Paleocene beds on the flanks of some uplifts. Basins between uplifted ranges in Wyoming, Colorado, and Utah became filled with several kilometres of stream and lake sediments, as shown on figure 13.2. Many small Tertiary intrusive masses, scattered from Mexico to Alaska, are also indicated on the map. Several Cenozoic intrusions were accompanied by copper, lead, zinc, and iron mineralization, which forms the basis of much western mining.

One of the most interesting and economically important Early Tertiary basin fillings is the Green River formation of Eocene age in Wyoming, Utah, and Colorado. Green River beds were deposited in a large lake, whose maximum extent is shown on figure 13.8. At certain times

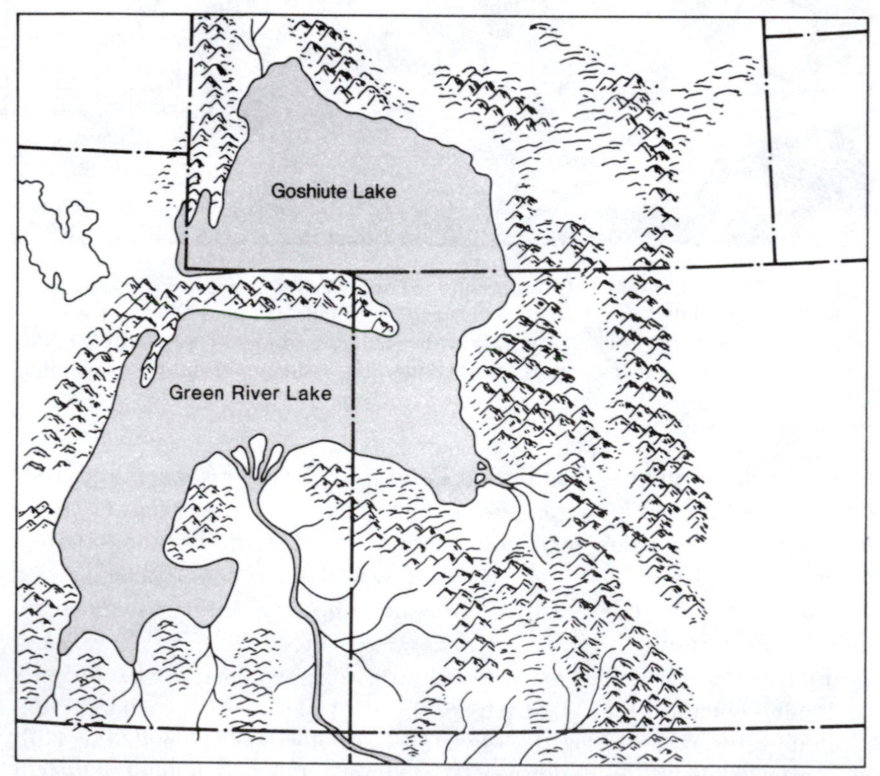

Figure 13.8. Eocene lakes in which Green River oil shale and related beds were deposited. These deposits form one of the few nonmarine petroleum reserves. The lakes occupied basins found in conjunction with Laramide uplifts.

the lake waters were very rich in microplants that were deposited in bottom muds and now form North America's largest oil shale deposit. At other times the lake waters evaporated and deposited trona, a sodium carbonate, which is mined in Wyoming. Uranium is also produced from these rocks in Wyoming. The lake abounded with Eocene life as indicated by fossil fishes, birds, crocodiles, and turtles, whose remains are common in Green River shale beds.

LATE TERTIARY OF THE WESTERN INTERIOR

Several major events occurred during Late Tertiary that have left a profound mark upon the deposits and topography of the western interior part of the continent.

1. Basin filling ceased in the Laramide intermontane basins (fig. 13.9). East of the Rocky Mountains a thin apron of Pliocene continental sediments was spread eastward onto the plains as shown on figure 13.3.
2. Block faulting, extending from the Nevada-Oregon border southeastward to central Mexico, created the late Tertiary "basin and range" topography consisting of alternating mountains and valleys. Normal faulting began in Miocene time and has been intermittenly active to the present. As the mountains were uplifted they were eroded, and their gravel, sand, and mud were deposited in the adjacent valleys, as shown on figures 13.11 and 13.12.
3. Great volumes of basalt were poured out of fissures and shield volcanoes in the Columbia River plateau of Washington, Oregon, and the Snake River plains of Idaho as shown on figure 13.3. These basalts are chiefly Miocene and Pliocene, but some vents were active within Pleistocene and Recent time, as exemplified by the Craters of the Moon area in Idaho.

   Along the west edge of the basalt plateau a great chain of andesitic strato-volcanoes extends from Mt. Lassen, California, northward. These were most active in Pliocene and Pleistocene time, but some, for example, Mount Baker in Oregon, have been mildly active up to the present. Smaller volcanoes of Late Tertiary age, chiefly basaltic cinder cones, are widely scattered in western North America and include such features as the San Francisco peaks in Arizona.

   Volcanic eruptions in western America reached a high point of activity during the Oligocene and Miocene. From Mexico to Alaska enormous amounts of volcanic material, mainly ash flow tuffs, were

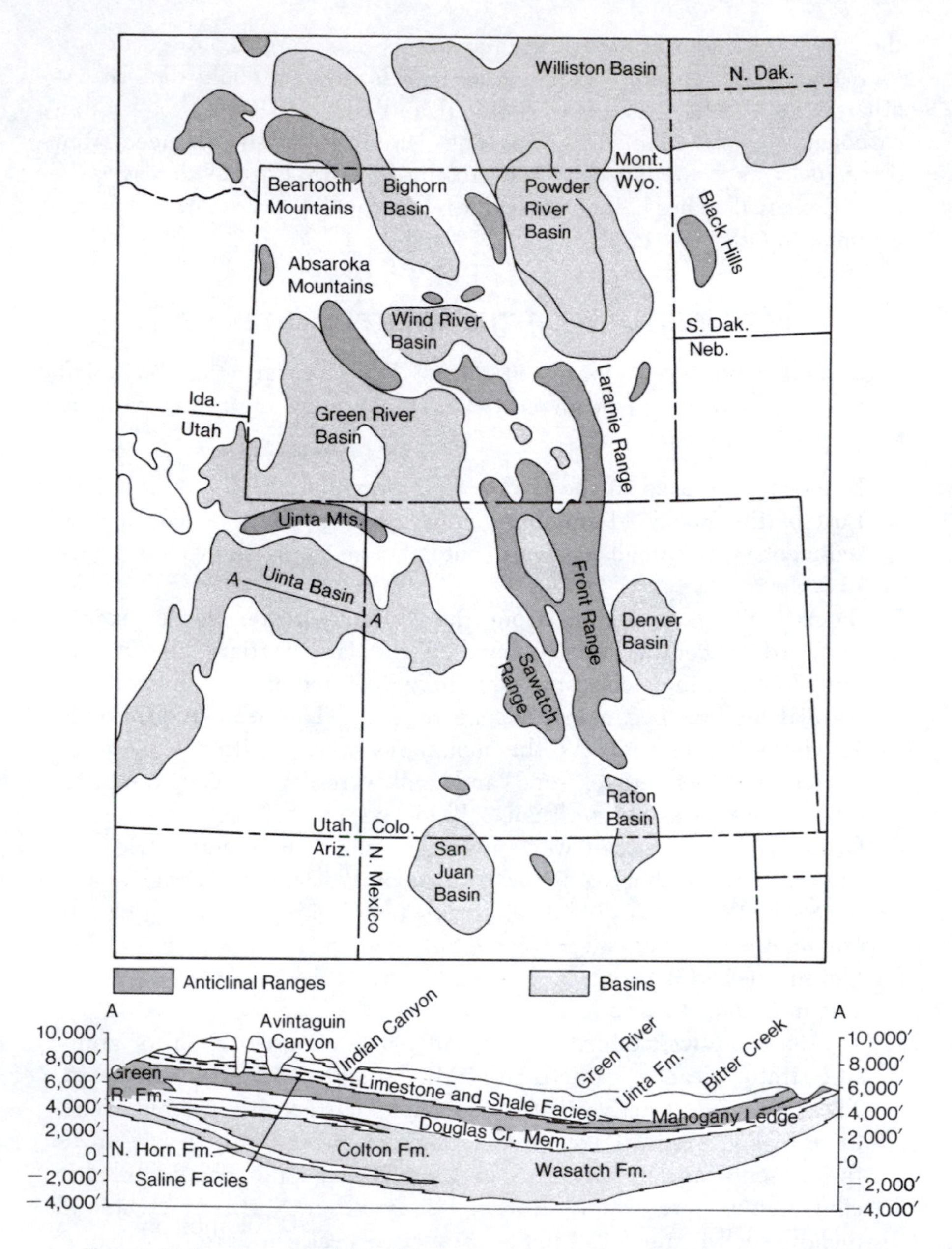

Figure 13.9. Map of Early Cenozoic mountains and basins in Western America with cross section through Uinta Basin in Eastern Utah.

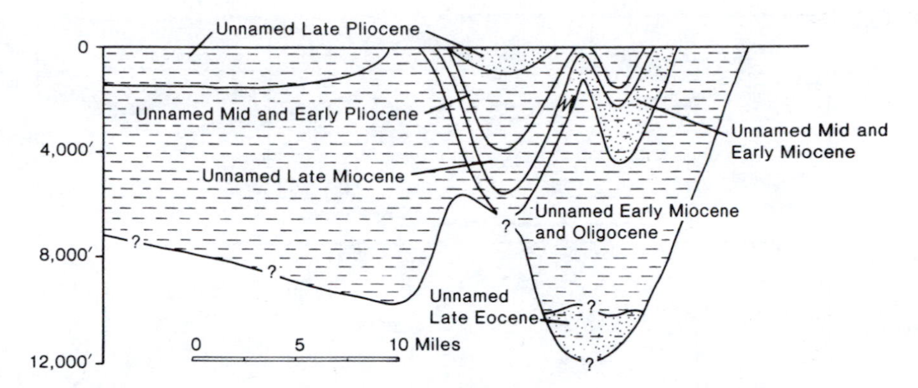

Figure 13.10. Cross section of Tertiary strata, offshore Vancouver Island, B. C. (From Shell Strat. Atlas of North and Central Amer., Cook, T. D., and Bally, A. W., editors, 1975, p. 270)

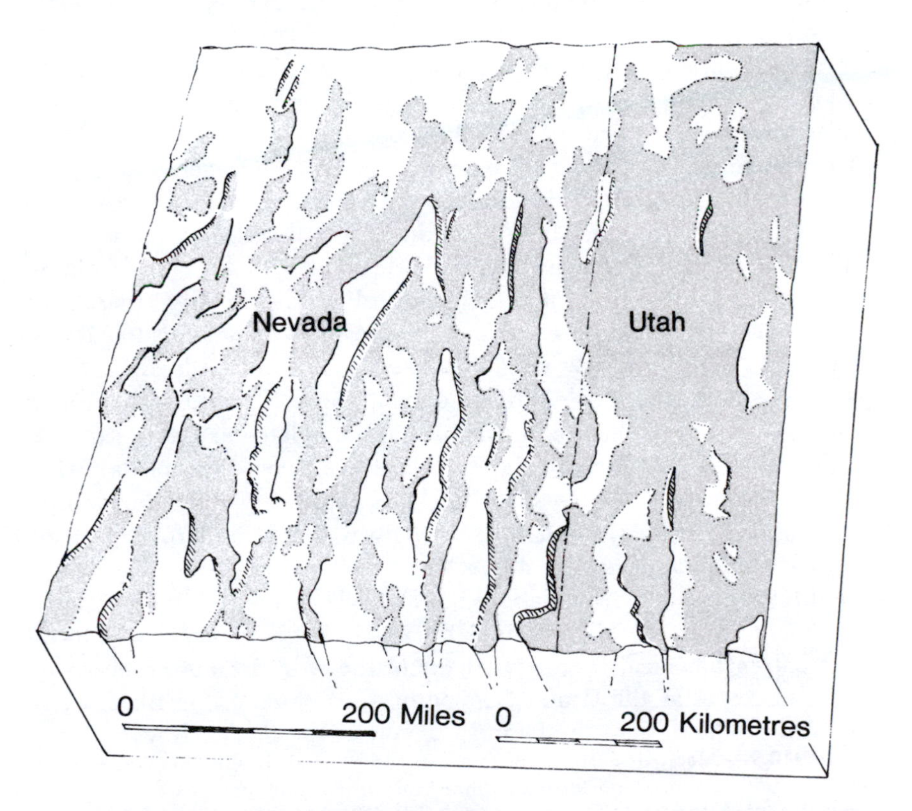

Figure 13.11. Late Cenozoic fault block basins of Utah and Nevada. Several thousand metres of displacement has occurred along many faults resulting in down dropping of valleys and upraising of mountain blocks.

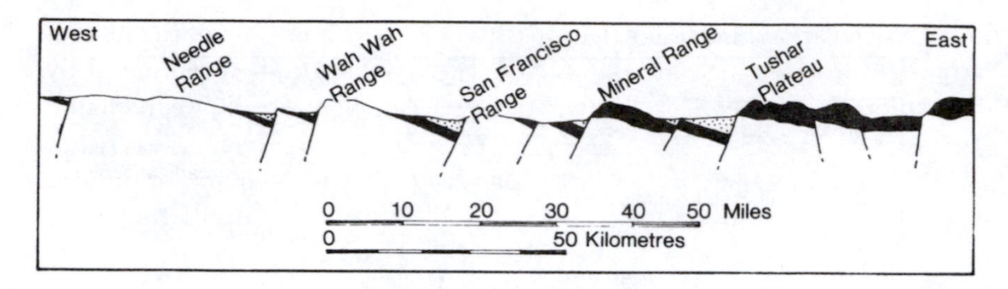

Figure 13.12. Cross section through part of the Basin and Range in southwestern Utah showing fault block relationships.

erupted during these epochs. The volcanic rocks yield radiometric potassium-argon dates of between 8 and 37 million years. The abundant volcanism in western America is interpreted to indicate the continued presence of a subduction zone along the Pacific Coast through Middle Tertiary.

4. The final episode of compressive folding occurred during the Miocene along the Pacific Coast, folding the Coast Ranges and Cascade Range from California to British Columbia. The orogeny is called by either of two names, Coast Range Orogeny or Cascadian Orogeny. In several places minor overthrusting occurred, not as extensive and forceful as in the Sevier Orogeny, but none the less impressive. Figure 13.10 illustrates folded Tertiary rocks at Vancouver Island, British Columbia. Middle Miocene strata are involved in the folding; therefore, the lower limit on the age of the compressive event is Middle Miocene. Note that the overlying Pliocene beds are not folded, but are separated from those below by an unconformity. Therefore, the upper limit on the age of the folding can properly be dated as Late Miocene. The Cascadian Orogeny of the American West coincides in time with the formation of the Alps and Himalayas in other parts of the world.
5. At the time the basin and range area was block faulted, the entire western interior of the continent from Denver to Reno was slowly uplifted several kilometres above sea level. This resulted in the rejuvenation of the Rocky Mountains to their present grandeur, as well as in the entrenchment of the major rivers of the west, forming such spectacular gorges as the Grand Canyon of the Colorado, the Black Canyon of the Gunnison, the Grand Canyon of the Snake, Royal Gorge, and Flaming Gorge.

By Early Miocene time the Laramide Rocky Mountains had been bevelled and adjacent basins filled leaving western North America much lower in elevation and relief than it is today. Its present magnificent topography is in large measure due to late Tertiary uplifting, accentuated by excavation and erosion of Early Tertiary basin deposits by rejuvenated streams.

At this same time in geologic history, the Appalachian Mountains were also being uplifted and rejuvenated. It is from this uplift that their modern erosion patterns stem. Uplifts of the two widely separated mountain regions, though at the same time, were responding to different forces. Uplift of the west was due to tectonism, whereas that in the east was due to isostatic compensation of the crust following the Mesozoic break up of Pangaea.

TERTIARY OF THE PACIFIC COAST

In contrast with the broad continental shelf along eastern North America, the western margin of the continent has an extremely narrow shelf. Sediments carried to the Pacific Ocean by the major rivers are deposited directly in deep water. Early Tertiary deposits exposed in the Coast Ranges of Washington, Oregon, and California contain tremendous thicknesses of marine strata consisting chiefly of poorly sorted sandstones and shales interbedded with submarine basalts, indicating deposition in deep near-shore waters.

In late Cenozoic time the Farallon plate of figure 13.2 was apparently consumed along the Middle American trench to such an extent that its remnants are shown on figure 13.3 as the Cocos, Rivera, and Juan de Fuca plates. In the process, the East Pacific Rise impinged against the North American plate and Baja California split off the North American plate, attached itself to the Pacafic plate, and was moved northward. Interaction between the Pacific and North American plates along southern California has created a series of active horizontally moving blocks, as shown on figure 13.14. Sediments have been deposited in inter-block basins in southern California and are very thick in local basins, such as the Los Angeles Basin and San Joaquin Valley, as shown on figure 13.13. Relationships are extremely complex in this mobile area of plate interaction. At present, the Pacific plate is moving northward past the North American plate at the rate of six centimetres per year. This motion not only explains the frequency of modern earthquakes in southern California but also is related to the complex Cenozoic history not only of California but the entire western United States.

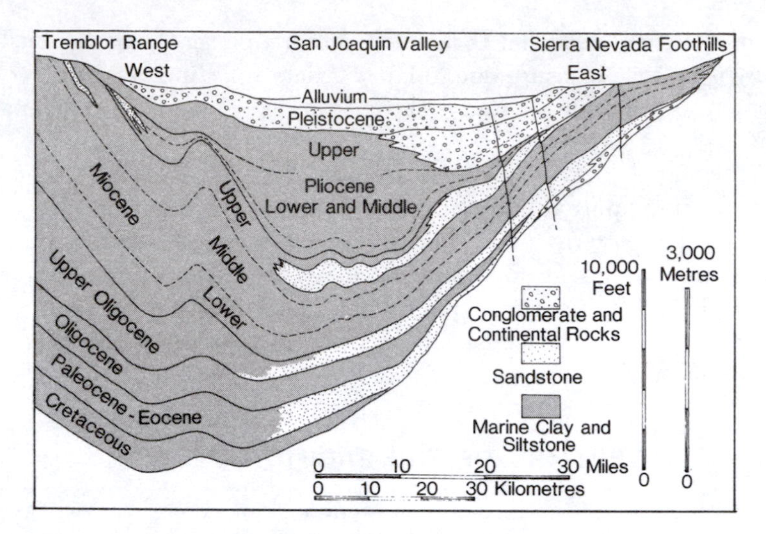

Figure 13.13. Tertiary rocks in the San Joaquin Valley near Bakersfield, California (modified from Hill, M. L. and Eckis, R., Calif. Dept. Nat. Res., Div. Mines Bull., v. 118, fig. 106, 1943). Oil occurs in Tertiary rocks where trapped by folds or unconformities.

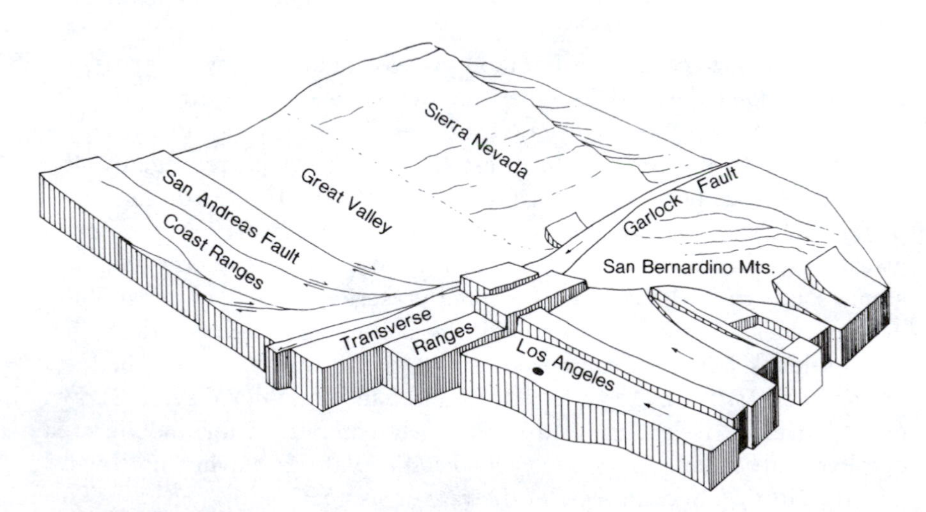

Figure 13.14. Schematic diagram of the area near Los Angeles showing how many small blocks have been raised, lowered or moved sideways along the San Andreas fault zone. The San Andreas faut zone is activated by movement between the Pacific and North American plates.

TERTIARY LIFE

Among the invertebrates of the Tertiary, microscopic foraminifera dominate the marine environment. Molluscs, specifically bivalves and gastropods, are the most abundant megascopic fossils. All have an obvious modern aspect, even those that lived early in the period. Tertiary foraminifera are perhaps the most intensely studied types of fossils because of their ubiquitous occurrence and consequent stratigraphic value in correlating Tertiary sediments, which produce more of the world's oil than rocks of any other period. One type of foraminifera common to Europe, Africa, and Asia, the nummultids, grew to the approximate size and proportions of a twenty-five-cent coin. The genus *Nummulites* is extremely abundant in the Eocene building stone from which the great Pyramid of Gizah, as well as other structures, were constructed.

Middle Tertiary marks the rapid development of the grasses which, in turn, greatly influenced mammalian evolution from mainly browsers to grazers. Like invertebrates, Tertiary plant fossils have a strikingly modern appearance.

The most characteristic forms of Tertiary life were the mammals. As reptiles dominated the Mesozoic, so mammals dominated vertebrate life during the Cenozoic. From humble beginnings in the Early Mesozoic, mammals evolved in such numbers and kinds that the Cenozoic is called the "Age of Mammals" (figure 13.15). With the extinction of predacious reptiles, mammals rapidly evolved, becoming adapted to fill most of the environmental niches recently vacated by reptiles. Mammals evolved to extremes in size, represented by the miniature shrew and the largest creature that has ever lived on earth, the modern Blue Whale. By Eocene time, early ancestors of modern horses, rhinoceroses, camels, rodents, bats, primates, and others were represented.

Oligocene mammals attained gigantic proportions. One group, the titanotheres, became the largest land animals on the North American continent, standing 2 metres at the shoulder while on all fours.

During the Early Tertiary North American mammals competed with those of Europe by migration across a land bridge. A Pliocene land connection between North and South America, which had not existed prior to that time, is indicated by the intermixing of North American mammals with those of South America, which resulted in the extinction of some less well adapted South American forms. Tertiary mammals competed among themselves; some were eliminated and others refined to inherit the Quaternary and the modern world. Increase of brain size, specialization of teeth to better accommodate a particular diet, and limb speciali-

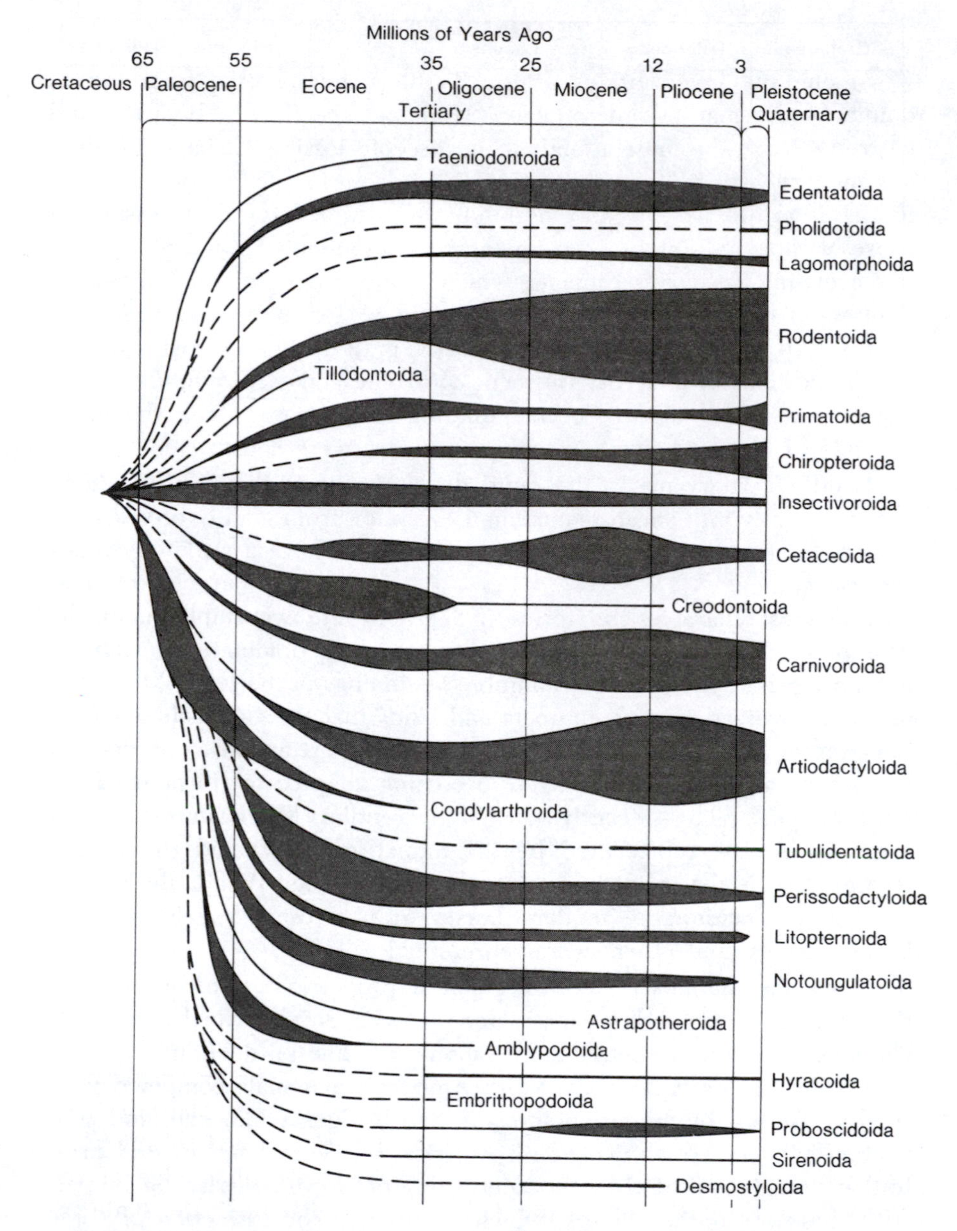

Figure 13.15. Chart showing origin and diversity of mammals during Late Cretaceous and Cenozoic time. (After Kurten, B., Scientific American, March 1969)

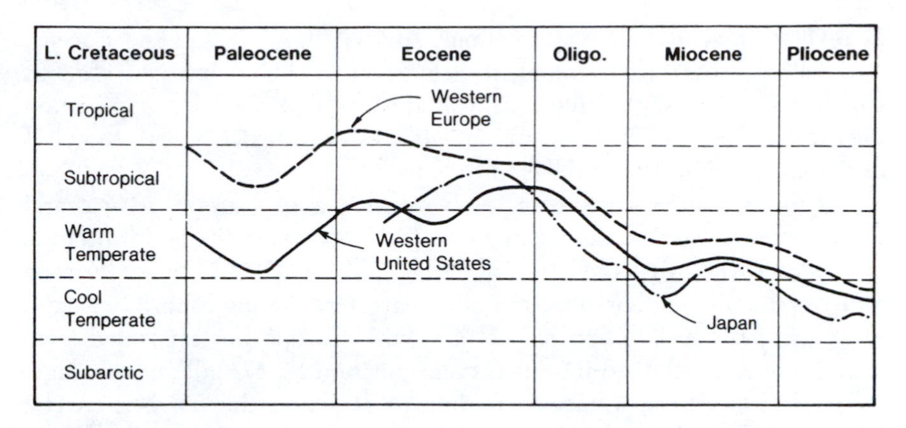

Figure 13.16. Average temperatures through Cenozoic time as inferred from fossil plants. (From Flint, R. F., Glacial & Quaternary Geology, John Wiley & Sons, Inc., p. 422, 1971)

zation to better equip the animal for life in a particular environment were the evolutionary trends of Tertiary mammals, as exemplified by the horse.

An almost stright-line decline in climatic temperatures on the continents occurred during the Tertiary. Figure 13.16 illustrates this general "cooling off" and accompanying shifts in Tertiary vegetation. A general decline in temperature during the Tertiary ushered in the Pleistocene, or Ice Age. This severe temperature change in the environment was a major pressure selecting the survivors of the Quaternary.

TERTIARY TECTONICS

The east and west coasts of North America behaved in markedly contrasting fashion during Cenozoic time in response to their differing relationships with boundaries of the North American Plate. The North American continent is attached to and moving with the floor of the Atlantic Ocean west of the Mid-Atlantic Ridge. Therefore, coastal plain deposits on the Atlantic and Gulf seaboards have not been deformed by any plate interaction. But western North America is on the leading edge of the North American Plate as it impinges against the Pacific Plate. Plate interactions along the west coast have given rise to the complex patterns shown on the western half of map figures 13.2 and 13.3. Figure 13.2 shows the restored position of spreading centers in the Atlantic and Pacific Oceans as of 40 million years ago, determined by extrapolating in reverse from present plate motions. At that time, during the Eocene, it is believed that a consuming trench extended along most of western North America and that the Pacific Plate was just about to impinge on the North American Plate at a point off western Mexico where the Gulf of California would later develop.

In the Atlantic area, the successive positions of Europe, Africa, and Greenland are shown on the map figure 12.18. Although the South Atlantic opened earlier during the Mesozoic, nearly all the drift between Europe and North America has occurred since late Cretaceous time. Greenland moved initially in late Cretaceous time with Europe but separated from it and moved northward into its present position by 47 million years ago, in the Eocene. Early Cenozoic folding of strata in the northern Arctic Islands of Canada probably reflects the movement of Greenland into position against the top of North America.

Folding of the Coast Ranges during mid-Tertiary seems to be a response to compression related to subduction along the western margin of North America. Late in the Tertiary Period, beginning perhaps 40 million years ago, the American continent overrode the juncture between the American and East Pacific plates, as well as part of the East Pacific Rise, as shown in figure 13.2 and 13.3. The convergent juncture between the American and East Pacific plates is well known along the west coast of Central America and South America (fig. 13.3). The axis of the East Pacific Rise bisects the Gulf of California and is presently opening that body of water. The Rise is again seen off the northwest coast of the United States. The connection between these two spreading centers is explained by one or both of the following possibilities.

1. The horizontal San Andreas fault is a transform fault offsetting the spreading centers in a north-south shear motion between the American and East Pacific plates. Movement on the San Andreas fault is on the order of one centimetre per year.
2. The concentration and alignment of earthquake epicenters from the end of the Gulf of California northward through the Basin and Range province in the western United States coincides with an area of high-heat flow, a measure of the earth's internal heat lost at the surface, and crustal thinning. Normal faulting, which characterizes Basin and Range structure, is extensional and may reflect a buried, perhaps inactive, spreading center beneath the crust.

Basin and Range faulting has been described as a result of shearing related to motion along the San Andreas fault. The age of the earliest normal faulting is thought to be Miocene, which corresponds in time to the first shearing motion of the San Andreas fault system. The consuming subduction zone between the mid-Atlantic and East Pacific plates, except where seen off Central America and Alaska, is conjectural if present at all. Within the plate tectonic model, motion is relative, allowing the migration of rift systems as well as lithospheric plates.

14

Quaternary

TOPICS

The final interval of earth history, the Quaternary Period, about 2 million years in duration, consists of two epochs, the Pleistocene, which ended approximately 10,000 years ago, and the Recent, or Holocene. Because Quaternary deposits are at the surface, our understanding of the period is more complete than that of any other. Present-day geologic processes and events continue to form the record of the Quaternary as the continents, coastlines, and ocean basins are modified.

QUATERNARY GEOCHRONOLOGY

Although many methods used in dating older materials are useful in studying Quaternary deposits, some techniques are unique to this period:

1. Dendrochronology, the study of annular growth rings in trees, has developed a chronological record that goes back several thousand years. Variations in climatic conditions affect most of the trees in a given area in about the same way, resulting in thick rings in wet years and thin ones during dry. Age relationships can be established by counting pairs of rings, thus the age of the tree, and comparing or correlating the tree-ring patterns with other trees in the area. Materials and their associated Quaternary events can sometimes be cor-

related with fossilized trees and thus dated by the standard growth patterns that have been established. Bristlecone pines, the oldest living organisms, found in Nevada and adjacent states, allow the extension of dendrochronology records about 7,000 years back from the present into the early Recent.

2. Varves are paired deposits of clay and silt that have accumulated on Quaternary lake beds. Each complete varve pair may represent an annual increment of sediment accumulation. It is thought that the coarser silt layer represents summer deposition, while the clay and some organic material settled out during freeze-over of the lake in winter. Like the patterns of tree growth, varves display regional weather-reflecting patterns, principally of layer thickness, which can be quantitatively correlated and used to construct a standard sequence. By including older and younger varve accumulations that can be tied into the standard section, the useful duration for varve geochronology is extended. It is estimated that varve chronology is useful to approximately 12,000 years back. Like dendrochronology, varve can be used to estimate the ages of glacial deposits as well as nonglacial Pleistocene sediments.
3. Radiometric dating techniques using carbon 14, potassium-argon, and uranium decay are also valuable in establishing age relationship of Quaternary materials. Carbon 14 is useful only for approximately the last 30,000 years. Pleistocene dates in the 30,000 to 2 million year range, too long for carbon 14 and too short for long half-life yardsticks like uranium 238, are measured in an ingenious way using ratios of protactinium 231 and thorium 230, both intermediate daughter products of uranium decay, and by using potassium-argon ratios.

CONTINENTAL ICE DEPOSITS

Figure 14.1 illustrates the Pleistocene ice sheet that covered North America. Beginning approximately 2 million years ago, two great ice masses began building up upon the continent, the Laurentide ice sheet in eastern Canada, and the Cordilleran glacial complex in western North America. The two ice masses combined at times to cover the northern part of North America from the Atlantic to the Pacific, covering the entire area indicated on figure 14.1 with a curious exception, the driftless area of southwest Wisconsin, illustrated on figure 14.3. Parts of Alaska and the Arctic coast also escaped glaciation. The thickness of ice in the Laurentide mass has been estimated to be over 2 kilometres. This roughly

Figure 14.1. Map of North America showing the extent of glacial ice covering the continent during the Pleistocene, and the two main centers of glacial expansion.

corresponds with known thickness of the icecaps of Greenland and Antarctica, which measure in excess of 3 kilometres.

As ice masses, or continental ice sheets, moved across the landscape, they scoured the soil and rocks in some places, while depositing their sedimentary load in others. Much of the topographic relief throughout the glaciated area of the content reflects the cut-and-fill action of glaciation.

LAURENTIDE ICE SHEETS

The enormous Laurentide ice mass covered most of northern North America (fig. 14.1). Extent of the glacial ice is well marked by the rock debris strewn upon the present-day surface. Glacial drift deposits consisting mainly of moraines, drumlins, kettles, and eskers are common throughout the area. In Canada the surface of the ancient shield is noticeably scratched and gouged, preserving the effects of rock-tools frozen into the base of the south- and westward-moving ice mass.

Why these glaciers formed is not clear. It has been estimated that a 3.3°C decrease in mean summer temperature in the Labrador-Ungava region would provide conditions necessary to allow an ice buildup of 2 kilometres thickness in only 20,000 years. Conditions necessary to initiate ice buildup are heavy snowfall during the winter followed by cool, short summers. The winter precipitation in this condition exceeds the summer melting and accumulation occurs. Given sufficient time, this imbalance would cause an ice buildup followed by outward flow that would, in turn, affect climatic modification, thus furthering the glacial expansion.

Continental glaciation during the Pleistocene was cyclic. Glacial advances are recorded by deposits of drift spread across the landscape (fig. 14.2). When the ice front retreats, the drift sheet is exposed to weathering and erosion that convert the rock and mineral fragments to soil that may be capable of supporting plant growth. If the glacier again advances depositing a second layer of drift, a stratigraphic record is constructed containing two till sheets separated by an ancient soil layer. Pleistocene stratigraphy in the Upper Mississippi valley contains four till layers separated by three soil or weathered strata. This record is interpreted as representing four glacial advances named, from the oldest to the youngest, after the states in which the drift sheet is well displayed: Nebraskan, Kansan, Illinoian, and Wisconsin Stages (fig. 14.3). Similar strata in Europe also indicates four glacial advances during the Pleistocene. The retreat or melting of the Wisconsin ice sheet is illustrated on figure 14.4.

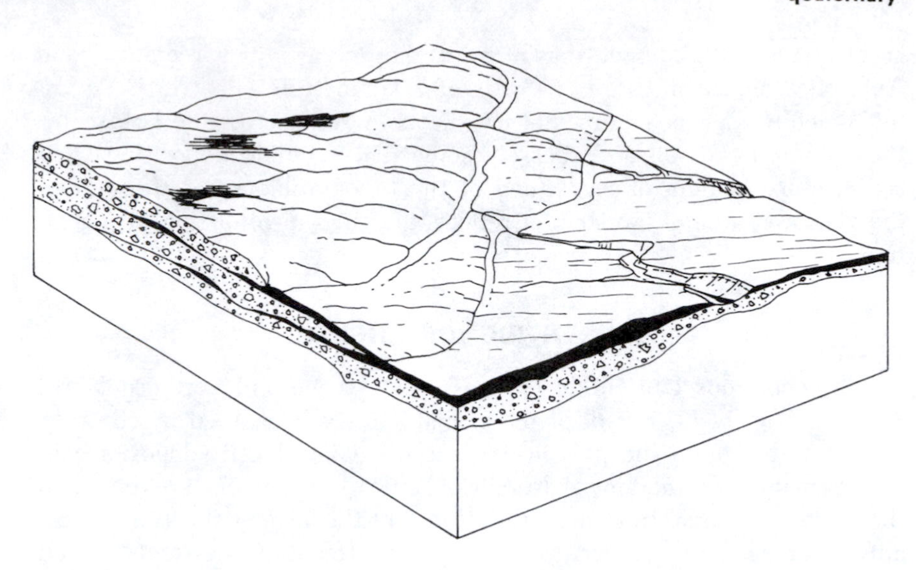

Figure 14.2. Diagram showing successive layers of till deposited by repeated ice advances. Dark areas indicate soil layers formed during interglacial periods.

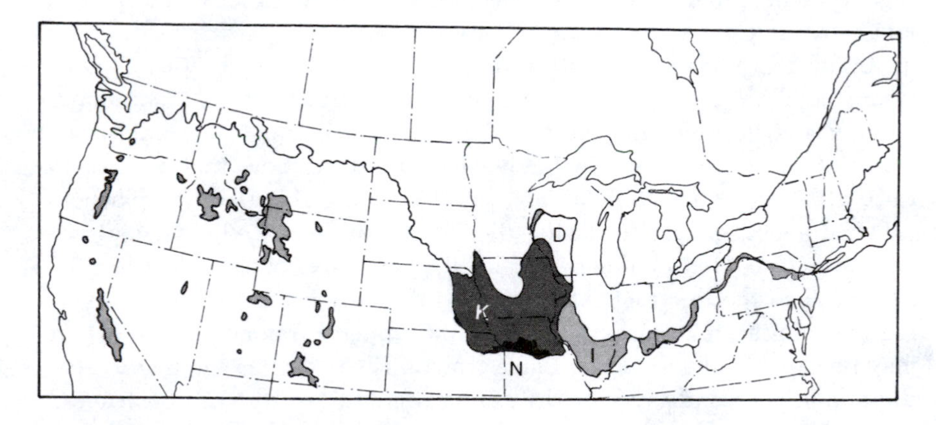

Figure 14.3. Map of the southern margin of Wisconsin ice sheet, showing relationships to earlier glacial stages. "N" stands for Nebraskan till, "K" for Kansan till, "I" for Illinoin till, and "D" for driftless area. (After Flint, R. F., Glacial and Quaternary Geology, John Wiley & Sons, Inc., p. 390, 1971)

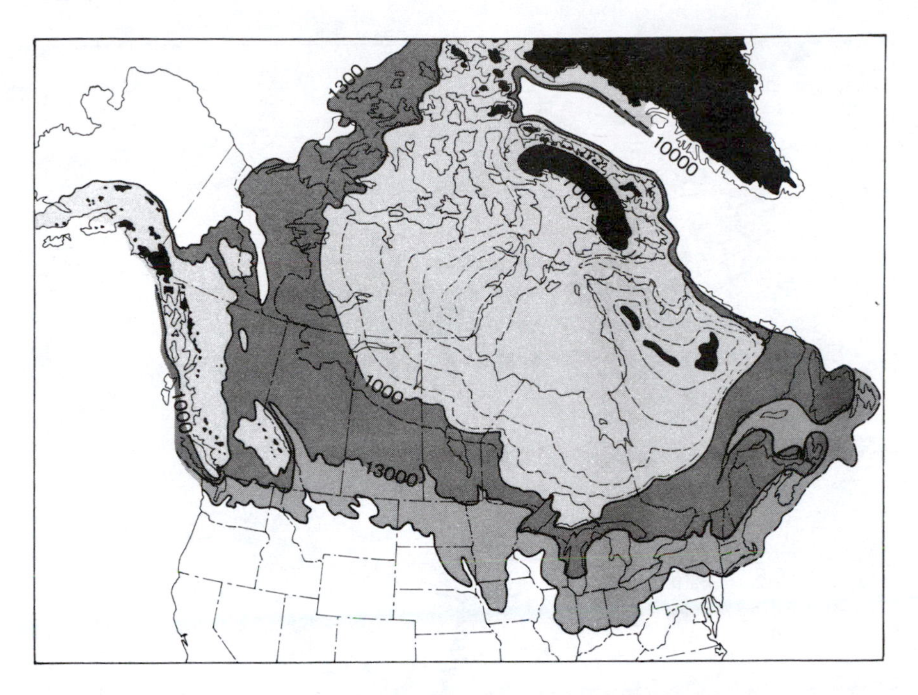

Figure 14.4. Map of North America showing deglaciation patterns of the Wisconsin ice sheet. Darker shades show progressively younger positions of the ice front. (After Prest, V. K., Quaternary Geology of Canada in Geology & Economic Minerals of Canada, p. 676-764, 1970) Figures indicate number of years ago.

CORDILLERAN GLACIAL COMPLEX

Cordilleran glaciers formed a continuous ice mass from the eastern Rockies to the Pacific coast extending from northern Washington to the Alutian Islands, a distince of nearly 4,000 kilometres. South of the main continuous ice mass were approximately 75 separate centers of glacial buildup, mainly centering on single high mountains, those above 3,000 metres, or a complex of mountains (fig. 14.1). Southern areas of greatest glacial development were the Yellowstone and Sierra Nevada mountains. High mountain peaks as far south as Mexico were glaciated during the Pleistocene.

Glaciers within the Cordilleran complex were alpine or valley type, frequently coalescing to form piedmont glaciers. Glaciers formed in the Cascade and Coast Ranges and Rocky Mountains because heavy precipitation resulted from orographic, or topographic, lifting of the eastward-moving, moist Pacific air masses.

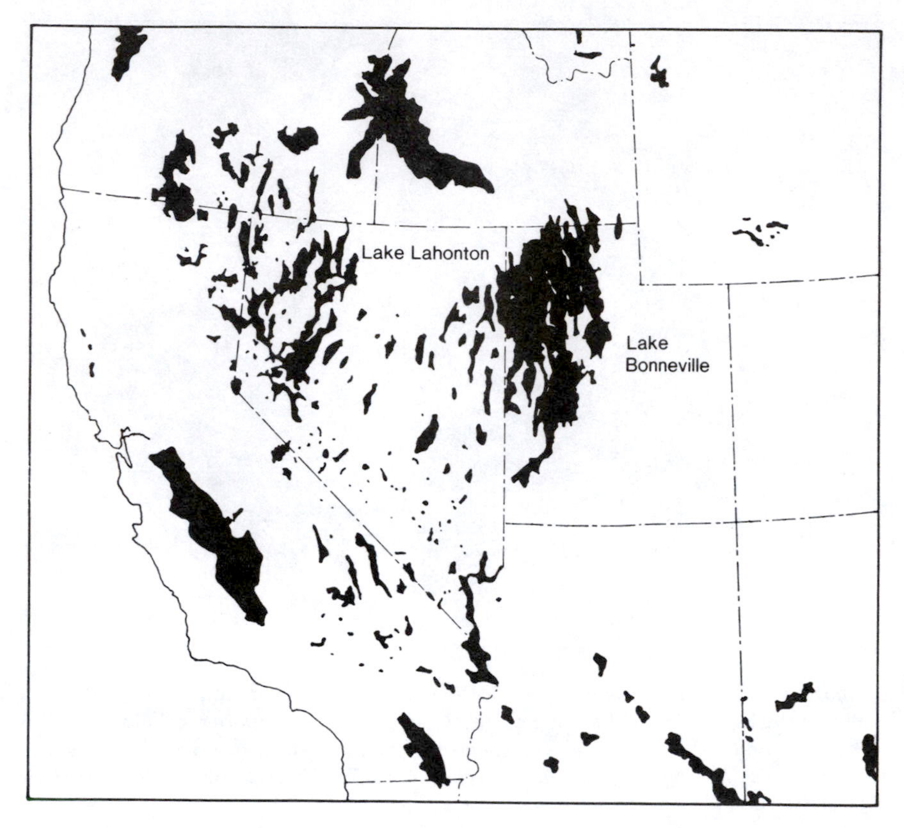

Figure 14.5. Map of western United States showing distribution of Pleistocene lakes whose waters were derived mainly from precipitation associated with the existing climates. (After Feth, J. H., U.S. Geol. Surv. Prof. Paper 424-B, p. 110-111, 1961)

The effects of Cordilleran glaciers are most noticeable in the development of erosional cirques, horns, arêtes, and U-shaped valleys. Typically associated with the more obvious erosional features are the glacial moraine deposits along the sides and termini of glaciated valleys.

PLEISTOCENE LAKES OF WESTERN UNITED STATES

The cool, moist climate of northern North America during parts of the Pleistocene was characterized by heavy precipitation – usually as snow northward and as rainfall in the south. In Utah and Nevada, two major lakes developed as rain and meltwater accumulated in the large basins. The erosional and depositional effects of Lake Bonneville in Utah and

Figure 14.6. Shoreline features of Lake Bonneville, in Utah. (From Gilbert, G. K., 1890, U.S. Geol. Surv. Mono. 1)

Lake Lahonton in Nevada remain beautifully preserved on the flanks of the present-day mountain ranges in these areas. At its maximum development, perhaps 18,000 years ago, Lake Bonneville was over 300 metres deep covering an area 250 kilometres by 550 kilometres (fig. 14.5). In some places, large beach sand and gravel deposits mark the ancient lake level, while at others prominent terraces extend for several kilometres, recording a temporary stationary position of the lake level (fig. 14.6). Isolated mountains, once protruding as islands, are encircled with terraces.

PLEISTOCENE RECORD IN OCEAN BASINS

Depletion of water from the oceans, as ice continued to pile up upon the continent, caused a relative lowering of sea level by an estimated 100 metres. With a lowered sea level, continental margins were exposed. North America was especially enlarged on its south and east margins where continental shelves are particularly wide. Across these newly exposed continental shelves rivers eroded channels into the Tertiary sediments. Known today as submarine canyons, they coincide with the deltas of the large rivers draining the present continents.

Sediments on modern ocean floors far from land are less than 200 metres in thickness. A record of climatic changes associated with the cyclic patterns of ice buildup and melting upon the continents is preserved within this rather thin veneer of sediments covering the moving sea floor.

The record is retrieved by taking deep-ocean core samples and studying the numbers, kinds, distribution, and growth patterns, particularly coiling direction of mainly planktonic foraminifera, now buried in sea-floor sediments. Oxygen-isotope ($0^{18}/0^{16}$) studies on sea-floor sediments reflect temperature fluctuation of the atmosphere, which, in turn, controlled the variables in planktonic foraminifera (fig. 14.7). A comparison of the separate approaches to paleotemperature studies reveal a similar pattern of cyclic temperature variation that correlates to ages of glacial

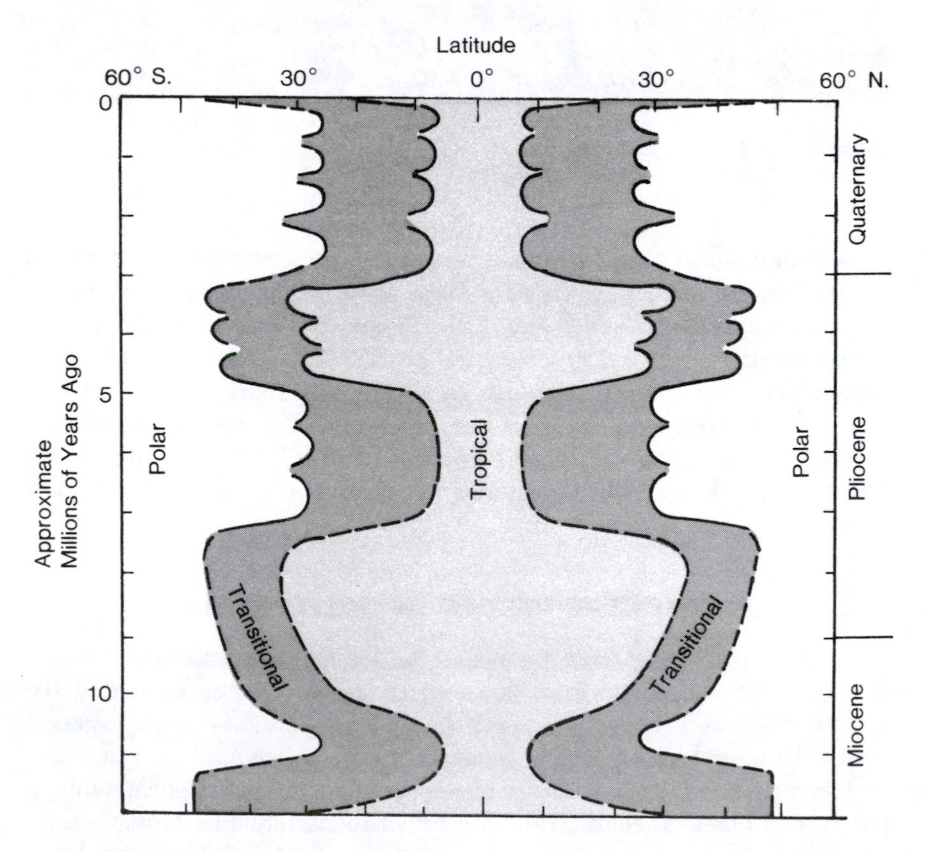

Figure 14.7. Diagram showing latitude displacements of planktonic foraminifera in the Pacific from the Miocene to the Recent. (After Bandy, O. L., Paleogeog., Paleoclim., and Paleoecol., v. 5, p. 63-75, 1968)

advances or retreats upon the continent, as illustrated on figures 14.9 and 14.10. As glaciers advanced, the cooling effect upon the atmosphere was reflected in the atmosphere of all latitudes, and the plankton in the sea compensated in several ways.

Figure 14.11 illustrates the development of the Mississippi River delta during the Quaternary. The river has altered its course across the expanding delta, depositing its sediment load at different points along the front of the delta. Higher numbers indicate progressively younger deposits.

ISOSTATIC REBOUND FOLLOWING THE PLEISTOCENE

As sediments are eroded, transported, and deposited upon the earth's surface, mass, represented by those sediments, is being moved from place to place. Where sedimentation occurs, such as in the geosynclines, additional mass is accumulated and the underlying rocks adjust by subsiding, whereas the site of erosion has been lightened and the strata rebound. This balance between earth materials of different density is called isostasy. We see isostasy in action in the melting of icebergs. They become smaller and smaller yet preserve the ratio 1:9 of ice above and below the surface of the sea.

During the Pleistocene, as mass (ice) was piled upon certain areas of the earth's surface, the crust responded by subsiding. As the ice melted, or as mass was removed, the crust rose or "rebounded." Crustal uplift due to isostatic rebound is measurable around Hudson's Bay, in Scandinavia, and within the basin of ancient Lake Bonneville in Utah, where the center of the basin has rebounded about 65 metres (fig. 14.8). Terraces that were constructed level around the lake are now several metres higher in the center of the basin where the crust subsided more deeply under the greatest load.

QUATERNARY LIFE

Planktonic foraminifera, bivalves, and gastropods are the most abundant marine forms. Modern ocean sediments consist in great part of the skeletal remains of these animals, for example, *Globigerina* and pteropod oozes that cover a large part of the sea floor.

Quaternary plants changed little since the Pliocene Epoch. Only a few species failed to adapt to the harsh Pleistocene climate. Most Quaternary plants adapted to the climatic change by migration. Their fossil distribution, known from pollen and macrofossils, reflects this adjustment to the changing climate. Subtropical floras existed in Washington and Ore-

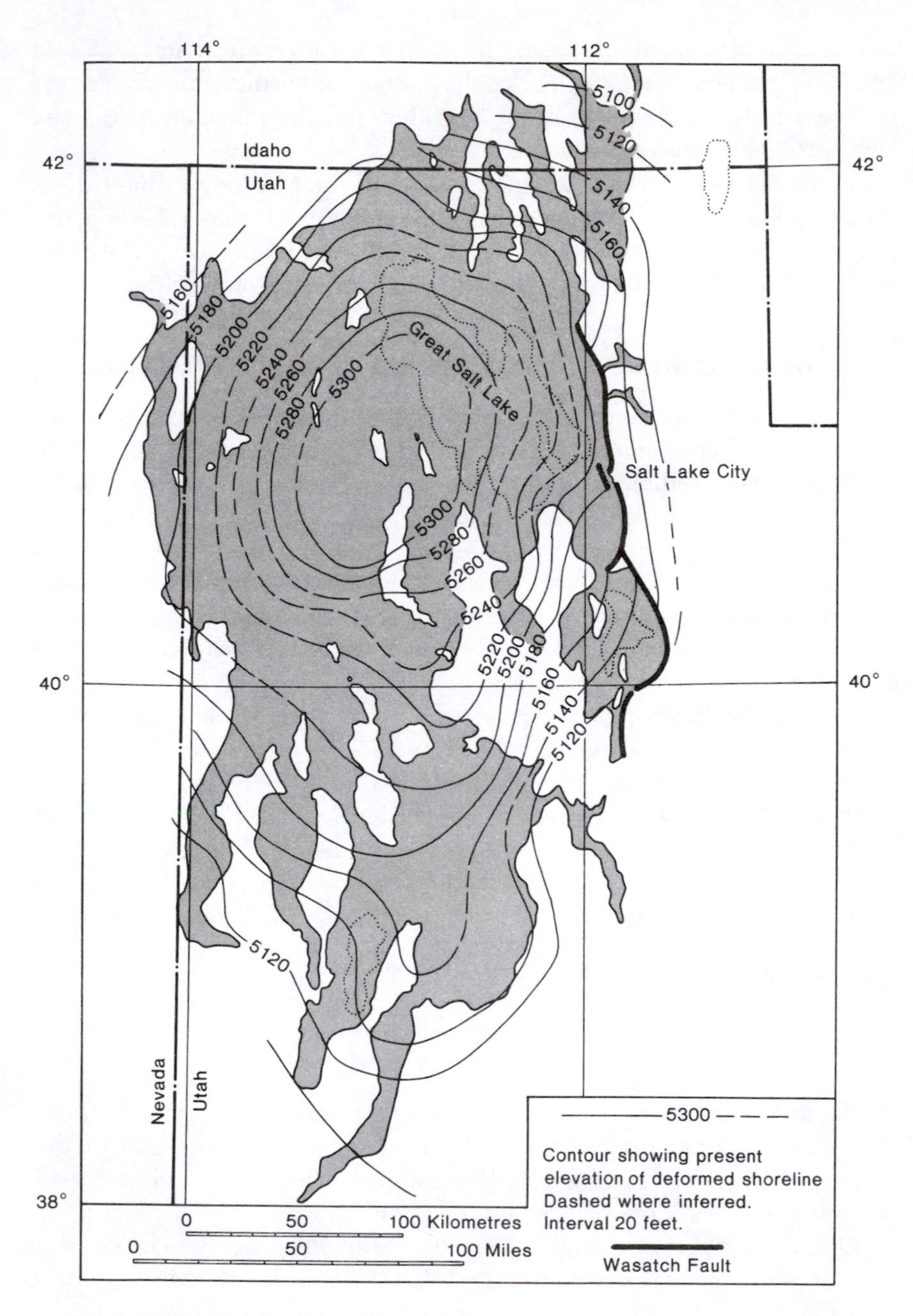

Figure 14.8. Contour map showing isostatic reboud of earth's crust following drainage of Lake Bonneville in Utah. (After Crittenden, M. D. Jr., 1963, U.S. Geol. Surv. Prof. Paper 454-E, p. E9)

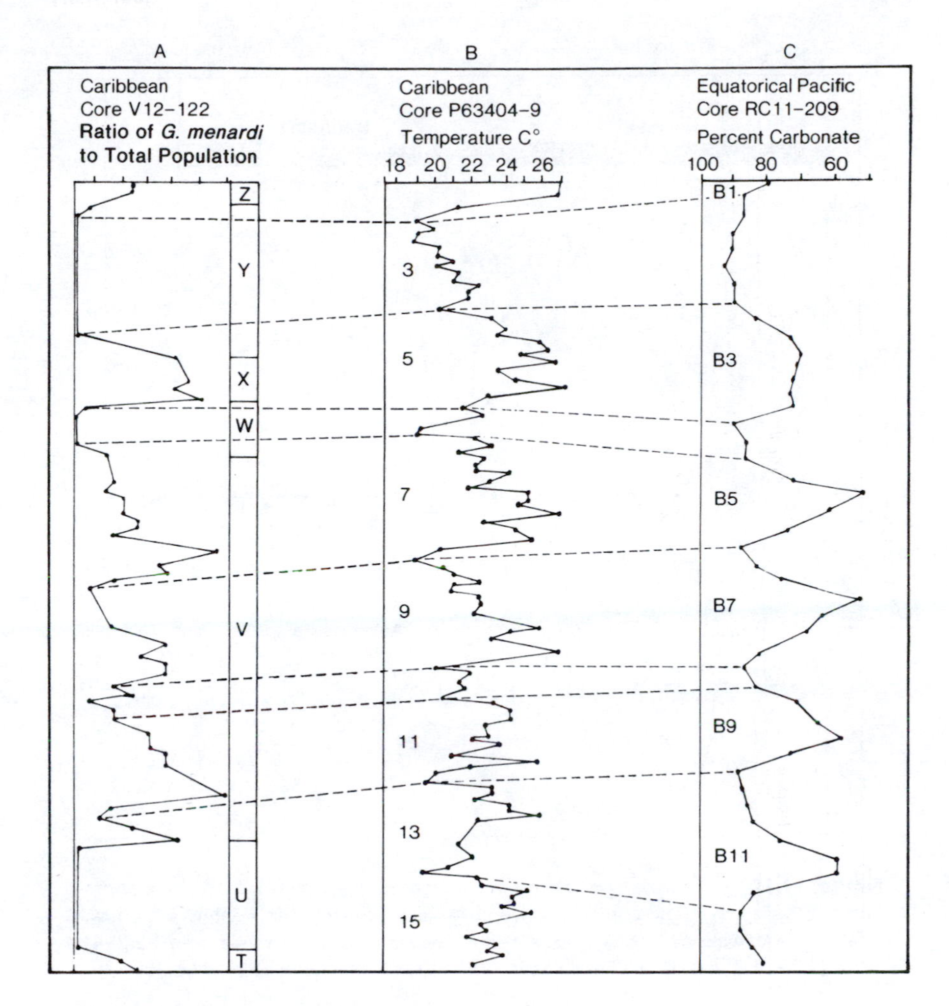

Figure 14.9. Diagram showing the correlation of (A) foraminifera temperature indications, (B) oxygen-isotope ratios and (C) foraminifera abundance indicated by carbonate percentage. The similarity of the three cores illustrates the worldwide effect on temperature of Pleistocene glaciation. (After Hayes, J. D., et al., Geol. Soc. Am. bull. v. 80, p. 1481-1514, 1969)

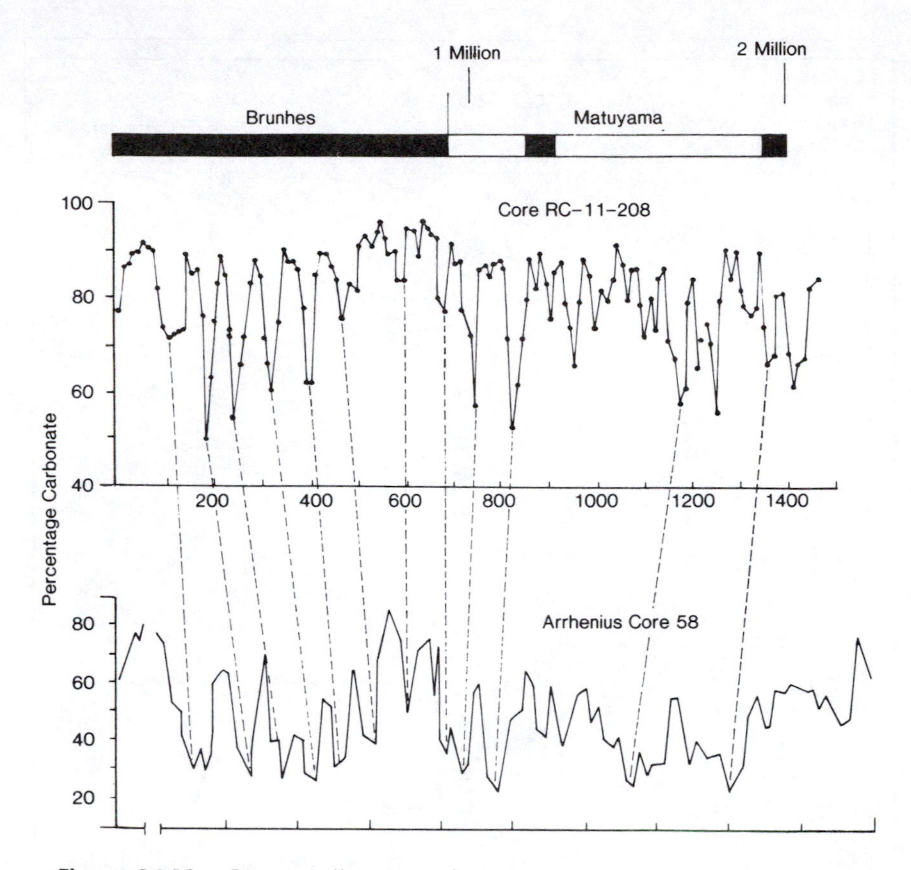

Figure 14.10. Diagram illustrating the correlation of two ocean-floor cores and the paleomagnetic conditions which existed during the past two million years. The calcium carbonate maximums reflect the abundance of foraminifera. (After Hayes, J. D., Geol. Soc. Am. bull. v. 80, p. 1481-1514, 1969)

gon as recently as Middle Miocene. As the ice-covered areas expanded, plants were driven southward.

Two general theories concerning plant survival during the Pleistocene are: (1) Isolated plant colonies continued to grow on nunataks, or mountain peaks projecting above the ice mass. This idea is supported by present-day isolated colonies of arctic plants in areas south of their normal range. (2) General retreat of all plant life before the expanding continental ice mass, presumably followed by a northward migration of the organisms following glacial retreat.

Early Pleistocene mammals of North America include horses, camels, beavers, mastodons, porcupines, deer, and an elephant-sized ground sloth.

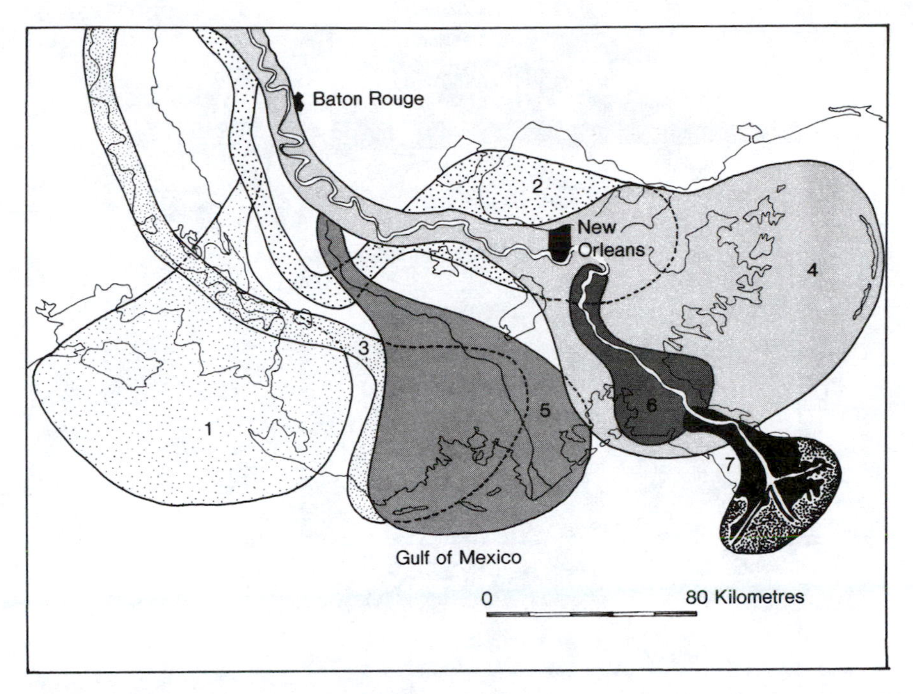

Figure 14.11. Diagram of the Mississippi River delta showing changes in its position during the Quaternary, numbered from oldest (1) to youngest (7). (After Fisk, H. N., et al. Miss. River Comm., 78 p.)

Mammoths, bears, stabbing cats, musk oxen, rodents, and bison followed soon afterwards. The wooly mammoth did not arrive in North America from its early home in Eurasia until the final glacial stage, the Wisconsin.

Tar pits dating back to Middle Pleistocene, called Rancho LaBrea for a locality in Los Angeles, have yielded numerous kinds of Pleistocene fossils. Bears, saber-tooth tigers, bovids, skunks, bats, birds, wolves, rodents, and mastodons have all been recovered from the tar.

Frozen within the Alaskan permafrost, skin and tissues of mammoths, mastodons, bison ,ground sloths, wolves, and many other mammals have been collected. Preservation of this type is essentially complete, providing information about the soft parts of animals, which is unavailable from other types of fossils.

Unlike the Quaternary plants, many large mammals became extinct late in the Pleistocene, approximately 11,000 years ago. Some authors estimate up to 70 percent extinction of those animals found in North America. The causes of the widespread extinction remains poorly under-

Figure 14.12. Drawing of Pleistocene mastodon whose fossil remains are found throughout much of North America. The wooly coat seems to be an obvious adaption to the cold climate that existed.

stood. Disease, man's influence, and climatic changes have been proposed to account for the rapid change in faunal makeup.

Among Quaternary mammals, man represents the most important addition. The oldest are, hominids, the australopithecines, are from South and East Africa, are perhaps as much as 2.5 million years old. From these oldest australopithecines presumably was derived *Homo erectus* and *Homo sapiens,* which appear in Middle and Upper Pleistocene sediments.

Early man's history in America is comparatively short compared with the record in Europe, Africa, and Asia. Few actual fossils of man have been found in North America. Only two occurrences of bones, one in Midland, Texas, and one at Tepexpan near Mexico City, are widely accepted, although artifacts from these early periods are more widespread. These are dated at approximately 10,000 years from the present. Carbon 14 dates of an early hunting culture in eastern Oregon and on the Great Plains are approximately 13,000 years from the present. It is generally thought that these early immigrants traversed a land bridge across the Bering Strait. Earlier migration across the Bering Strait southward would likely have been blocked by the Late Wisconsin ice sheet.

QUATERNARY TECTONICS

Modern tectonics is the basis for our interpretation of earlier earth history. Scientists are actively trying to confirm the plate tectonic model by actually measuring active plate motion. Following Hutton's principle of uniformity, we reason that many, though not necessarily all, processes that have been active in the past are with us today. Recent workers have suggested that geosynclines are now forming off the eastern coast of North America where today the sedimentary deposits at the base of the continental slope are approximately 10 kilometres thick composed of typical eugeosynclinal deposits lacking only the igneous rocks and metamorphism to duplicate Devonian eugeosynclinal strata. Upon the modern continental shelf, sediments total three to five kilometres in thickness and are characteristic of miogeosynclinal types. The future of these sediments rests in the pattern of plate motion of the American plate. If it were to change direction and duplicate its early Paleozoic motion, the sediments along the Atlantic coast would be compressed, intruded, and uplifted into another orogenic belt.

The modern Pacific coastline of Asia with the volcanic island-arc chains like Japan likely resembles the earliest stages of geosynclinal deformation similar to ancient coastlines of the American continent.

Present-day rift sysιems are separating the coasts and thus enlarging the Gulf of California and the Red Sea. Iceland is being pulled apart by active rift system beneath it.

Plate tectonics as a model for understanding earth history can be summarized as involving compression, intrusions, volcanism, and earthquakes along convergent plate margins. Divergent plates produce block-faulting, volcanism, and shallow seismic activity. Horizontal, or shear, motion between plates is seen by transform faults such as those along the rift axes within the ocean basins and faults like the San Andreas in California.

A FINAL LOOK BACK

Our approach to earth history has been conventional in that we have followed the "layer-cake" model of describing events and materials within each separate geologic period, an approach that has a built-in emphasis based upon the fossil record. Classifying the history of the earth into less traditional divisions, based upon the tectonic cycle, the following sequence could be used:

1. PRECONTINENT INTERVAL – the time from the beginning of the earth until the first permanent continents were formed. The rock record for this interval is not found on earth, but is well preserved on the Moon. Earliest earth rocks were recycled and consumed in the mantle until the continental crust of lighter granite rocks were formed. (4.65-3.7 b.y.)
2. ARCHEAN INTERVAL – containing the earth's oldest rocks, formed as the continents were first developing and becoming increasingly more silicic as geologic processes concentrated the lighter rocks nearer the surface. Mafic graywackes, conglomerates and volcanics were being formed. (3.7-2.4 b.y.)
3. PROTEROZOIC INTERVAL – the time of continental enlargement as shelves were built by accretion. Upon them the shallow seas washed and sorted the sediments, accumulating for the first time widespread clean limestone, dolomites, and iron-bearing chert beds. Upon these shelves early life evolved. (2.4-.6 b.y.)
4. CONSTRUCTION OF PANGAEA – an interval recording the building of Pangaea. An interval of relatively worldwide mountain building. (Cambrian-Devonian)
5. PANGAEA – an interval of widespread, relatively uniform conditions reflected in the widespread glaciation of the Permian Period and arid conditions of the Triassic. (Mississippian-Triassic)
6. BREAK-UP OF PANGAEA – the fragmentation of the supercontinent, which began in the late Triassic and continues today. (Triassic-Present)

Index